CARS

**Other titles in
this series include:**

COMPUTERS
FASHION
FITNESS
HEALTH CARE
MUSIC

CARS

by Michelle Krebs

Series developed by Peggy Schmidt

Peterson's

Princeton, New Jersey

A New Century Communications Book

Library of Congress Cataloging-in-Publication Data

Krebs, M. (Michelle)
    Cars / by Michelle Krebs.
        p.   cm.—(Careers without college)
    "A New Century Communications book."
    ISBN 1-56079-221-3 (pbk.) : $7.95
    1. Automobile industry and trade—United States—Vocational
guidance.   2. High school graduates—Employment—United States.
I. Title.   II. Series.
HD9710.U52K688   1992
629.2'023'73—dc20   92-28000
                                                     CIP

Art direction: Linda Huber
Cover photo: Bryce Flynn Photography
Cover and interior design: Greg Wozney Design, Inc.
Composition: Bookworks Plus
Printed in the United States of America
10 9 8 7 6 5 4

Text Photo Credits
Color photo graphics: J. Gerard Smith Photography
Page xviii: © Woodfin Camp & Associates, Inc./Michael L. Abramson
Page 18: © Stockphotos/David Perry Lawrence
Page 32: © The Image Bank/Bill Varie
Page 48: © Joel Gordon Photography
Page 66: © Bruce Curtis

ABOUT THIS SERIES

Careers without College is designed to help those who don't have a four-year college degree (and don't plan on getting one any time soon) find a career that fits their interests, talents and personalities. It's for you if you're about to choose your career—or if you're planning to change careers and don't want to invest a lot of time or money in more education or training, at least not right at the start.

Some of the jobs featured do require an associate degree; others only require on-the-job training that may take a year, some months or only a few weeks. In today's real world, with its increasingly competitive job market, you may want to eventually consider getting a two- or maybe a four-year college degree in order to move up in the world.

Careers without College has up-to-date information that comes from extensive interviews with experts in each field. It's fresh, it's exciting, and it's easy to read. Plus, each book gives you something unique: an insider look at the featured jobs through interviews with people who work in them now.

Peggy Schmidt

ACKNOWLEDGMENTS

Special thanks to the following people and organizations for their contributions to this book.

Catherine Bonadeo Ahles, Vice President for College Relations, Macomb Community College, Sterling Heights, Michigan

Alex Fedorak, Public Relations, Subaru of America Inc., Cherry Hill, New Jersey

Ruth Gastel, Public Relations, Insurance Information Institute, New York, New York

Paul Gould, Co-op Coordinator, Macomb Community College, Sterling Heights, Michigan

David Hederich, Public Relations, Chevrolet Motor Division, Warren, Michigan

Lee Iacocca, Chairman, Chrysler Corporation, Highland Park, Michigan

Beth Kapchonick, Director of Public Relations, Detroit Diesel Corp., Detroit, Michigan

Kevin Kennedy, Public Relations, Campbell and Company, Dearborn, Michigan

Linda Kosinski, Public Relations, General Motors Acceptance Corporation, Detroit, Michigan

Ed Kovalchick, President, Net Profit Inc., Birmingham, Alabama

John Krudy, Former Associate Dean of Technology, Macomb Community College, Sterling Heights, Michigan

Arthur C. Liebler, Vice President of Communications, Chrysler Corporation, Highland Park, Michigan

Terry Lynn, Senior Project Engineer, General Motors Corporation, Warren, Michigan

Gene McKinney, Director of Communications, Motor Vehicle Manufacturers Association, Detroit, Michigan

Ted Orme, Public Relations, National Automobile Dealers Association, McLean, Virginia

Alva Powell, Faculty, Mechanical Technology, Macomb Community College, Sterling Heights, Michigan

Mark Rollinson, Public Relations, Buick Motor Division, Flint, Michigan

David Sloane, Public Relations, General Motors Corporation, Detroit, Michigan

William Thompson, People Development Education Department, Progressive Insurance Company, Beechwood, Ohio

John Walmsley, Media Relations and Special Programs Administrator, USAA, San Antonio, Texas

John Waraniak, Engineer, Chevrolet Motor Division, Warren, Michigan

Patrick Wright, President, Patrick Wright Enterprises, Grosse Pointe, Michigan

Alliance of American Insurers, Schaumburg, Illinois

Automotive Service Industry Association, Elk Grove Village, Illinois

AutoWeek, Detroit, Michigan

National Institute for Automotive Service Excellence and the National Automotive Technicians Education Foundation, Herndon, Virginia

Society of Automotive Engineers, Warrendale, Pennsylvania

Society of Manufacturing Engineers, Dearborn, Michigan

Linda Peterson, for her editing expertise

WHAT'S IN THIS BOOK

WHY THESE CAR CAREERS?

The automobile industry is a kingpin in the economy of the United States. Just about every household in the U.S. owns at least one car. Some have two or three. On any given day, 123 million cars travel the nation's roadways. About 30 million times a year, U.S. consumers buy or lease a new or used car, truck or van. And perhaps most importantly, one in seven jobs in this country depends on the automotive industry.

This book features five careers in the auto industry. They include:

❑ CAD (computer-aided design) specialist

❑ Car salesperson

❑ Mechanic

❑ Claims representative

❑ Race car driver

Each was selected because it is a career that does not require a four-year college degree. In fact, while some require specialized training, only one requires even a two-year associate degree. With the exception of race car driver, job openings are predicted to be plentiful in these careers throughout this decade and beyond. The work of people in each of these careers is important to the future of the car business.

A CAD specialist is in on the earliest stages of the birth of an automobile. Using the most modern computer technology available, the CAD specialist works with engineers and designers to create the blueprint for the car and its parts. Demand for CAD specialists far exceeds the supply, the pay is good and the potential for advancement is excellent. Because the technology is new and ever changing, a CAD specialist faces exciting challenges.

Once a car is manufactured, someone must demonstrate and sell it. The role of the car salesperson in the dealership is a changing one. The basic job remains the same; but with the increased emphasis on customer satisfaction, the salesperson has become the dealership's goodwill ambassador as well. Increased attention is also being placed on the job satisfaction of the salesperson. Dealerships are experimenting with ways to reward and keep good sales personnel beyond the reward that's always been there—making a good salary from commissions.

No matter how well a car is manufactured, it's going to need servicing and repair over its lifetime. Today's mechanic must use a computer as well as nuts and bolts because electronics are increasingly controlling car systems. It's still a job that requires getting your hands dirty, but it's one that pays a good hourly rate and is recession proof.

With more cars on the road than ever, the chance of accidents increases; so do jobs for auto claims representatives. They are the people who decide how much money car owners who have been in accidents are entitled to receive. Jobs are available in just about every small town and major city across the country. And insurance companies offer claims reps good potential for advancement.

Car racing is becoming more important to the automobile business because races draw big crowds and big dollars from corporate sponsors. Racing also affords car manufacturers the opportunity to test the latest technology on the track. Only a select number of people who fantasize about becoming race car drivers will ever know the thrill of competing in races because doing so requires talent and the ability to raise money. But for those who are successful, the financial rewards are great.

If you are someone who is into cars, chances are good you will be able to find a place in the automotive industry that is right for you.

LEE A. IACOCCA

on Cars and Your Future

No one in the automotive industry is better known than Lee Iacocca. He is credited with saving Chrysler Corporation from bankruptcy. He became familiar in every household through Chrysler's famous television ads in which he declared: "If you can find a better car—buy it."

Earlier in his automotive career, he climbed the corporate ladder at Ford Motor Company eventually to become president, a job from which he says he was fired by Henry Ford II. At Ford, Iacocca introduced the phenomenally successful Ford Mustang into the market.

Outside of the auto industry, Iacocca headed the project to restore the Statue of Liberty and Ellis Island—symbols of the opportunity the United States offered his Italian immigrant parents. He is repeatedly mentioned as a possibility for public office, including that of the presidency.

Although he is retiring from his position as chairman of the board and chief executive officer of Chrysler Corporation at the end of 1992, he will continue to serve on the board of directors in a key role.

We asked him to talk about how he got started and to advise our readers about careers without college in the automotive industry.

When I was in college, I drove a beat-up 1938 Ford. It was only 60 horsepower. More than once I'd be going up a mountain when suddenly the cluster gear in my transmission would go. I figured somebody at Ford decided they could get better fuel economy by taking a V-8 engine down to only 60 horsepower. That was a good idea—if they didn't have to drive in the mountains of eastern Pennsylvania. I used to tell my friends that anybody who builds cars that bad could use my help. That's how I got interested in the auto industry.

In August 1946 I began working at Ford Motor Company as a student engineer. I was assigned to the River Rouge plant, which at the time was the largest manufacturing complex in the world. Ford owned the coal and limestone mines, so we got to see the entire process, from hauling the minerals out of the ground to making the steel and then turning the steel into cars.

I found I was more interested in people than machines. So I moved into sales and marketing. My first sales job was in Chester, Pennsylvania, speaking to fleet purchasing agents about the allocation of new cars. It wasn't easy. I was bashful and awkward; I used to get the jitters every time I picked up the phone. So I would practice my speech before I'd make the call.

Learning the skills of selling was the hardest part of my job in those days. Some people think that good salesmen are born and not made, but I had no natural talent. I got good by practicing over and over until it became second nature. Not all young people understand that. They look at a successful businessperson and they don't stop to think about all the mistakes that the person made when he or she was starting out.

All the hard work paid off. After holding a variety of sales and marketing jobs, including general manager of Ford division and being in charge of planning, production and marketing of all Ford, Lincoln and Mercury cars and trucks, I felt I had arrived in my career on December 10, 1970. That's the day I became president of Ford. I remember thinking this is the greatest Christmas present I've ever had. I called my wife. I called my father. This was one of the happiest moments in my life. When I became president, Ford had 432,000 employees. Worldwide, we were building close to 5 million cars and trucks a year. And we were making a ton of money.

Over the years I felt like throwing in the towel a lot of times, especially when I was fired from Ford and went to Chrysler, which was on the brink of going bankrupt. But I never quit, even during those dark days at Chrysler, because the livelihoods of 600,000 people depended on Chrysler making it. The people of Chrysler are what I love most about my job. They're a very proud bunch—and they're very good and talented. We don't have the financial resources Ford and General Motors have, so we have to make more use of our people resource.

I've had a great career in the auto industry. I've had some proud achievements. Bringing Chrysler back from near bankruptcy is the biggie. I'm proud of a lot of products we did—the Mustang at Ford, the K-car and minivan at Chrysler and the new products coming out of Chrysler in the 1990s. I'm proud that I had the opportunity to raise money to restore the Statue of Liberty and Ellis Island.

I've been getting up and coming to work every day for 46 years. And would you believe it? I still love coming in. I'll miss it when I retire. If you don't keep challenging yourself, you start to waste away and I don't want to do that.

This may sound corny, but I like my job because I have the power to make a difference. I know a lot of heads of institutions who think their lot in life is to protect what they're handed. What you're handed has nothing to do with the present. Your legacy should be that you made it better than it was when you got it. I think I've done that at Chrysler.

The auto industry is an exciting business. I think it's the most important industry in the country; one in seven jobs depends on it. The economy isn't strong unless the auto and housing industries are strong.

But today's challenges are immense. That means there are important jobs for people with and without college degrees. In the corporate offices, many of our secretaries and technicians are high school grads. There are a lot of plant jobs that don't require a four-year degree. The same is true in dealerships—mechanics, service and parts and sales people. The degree isn't so important as the skills and dedication you bring to a job. Hard work and common sense are what it takes to succeed.

No matter what business anyone goes into, the key is to study hard. You've got to turn off the stereo. And the radio. And the TV. Don't accept phone calls. Then bury yourself in books for three hours a night. When you're done, read a book for pleasure instead of switching on the TV.

The problems of the auto industry are different from the ones in the days when I drove my beat-up Ford. The industry has become a global one. You can have the greatest product, but if you don't have good public policy to go along with it, product quality doesn't matter. We still have plenty of problems in Detroit, so if there are any bright people who have good ideas, we need them.

FAMOUS BEGINNINGS

Mario Andretti, Race Car Driver

Andretti has dominated Indy car racing for 28 years. He is considered one of history's top race car drivers. He was the first, and is only one of two drivers to have captured both the Indy Car and Formula One World Championships. He also was the first person to win both the Daytona 500 and the Indianapolis 500. He has won four Indy Car championships (in 1965, 1966, 1969 and 1984). He won the 1969 Indianapolis 500, setting new speed records. His first job was parking cars with his brother, Aldo, though neither boy could legally drive.

Jackie Stewart, Broadcaster and Consultant to Ford Motor Company

Stewart was the most dominant driver of Formula One race cars from 1969 to his retirement in 1973. He won the world championship in 1969, 1971 and 1973. During his nine seasons in World Championship Formula One competition, he won 27 races of 99 starts, a record that stood for 14 years. His first job was car racing at age 23.

Roger S. Penske, Race Car Owner, Car Dealer, Industrialist

Penske is founder and president of Penske Corporation. Penske Racing is the most successful Indy car racing team in history with a record eight Indianapolis 500 victories. He owns and operates professional motorsports facilities in Michigan, Ohio and Pennsylvania. His automotive retail operations include Penske Cadillac in Downey, California, and Longo Toyota, in El Monte, California. He heads Detroit Diesel Corporation, which makes diesel engines, and he runs Penske Truck Leasing Company. His first job was a sales engineer with ALCOA and part-time race car driver.

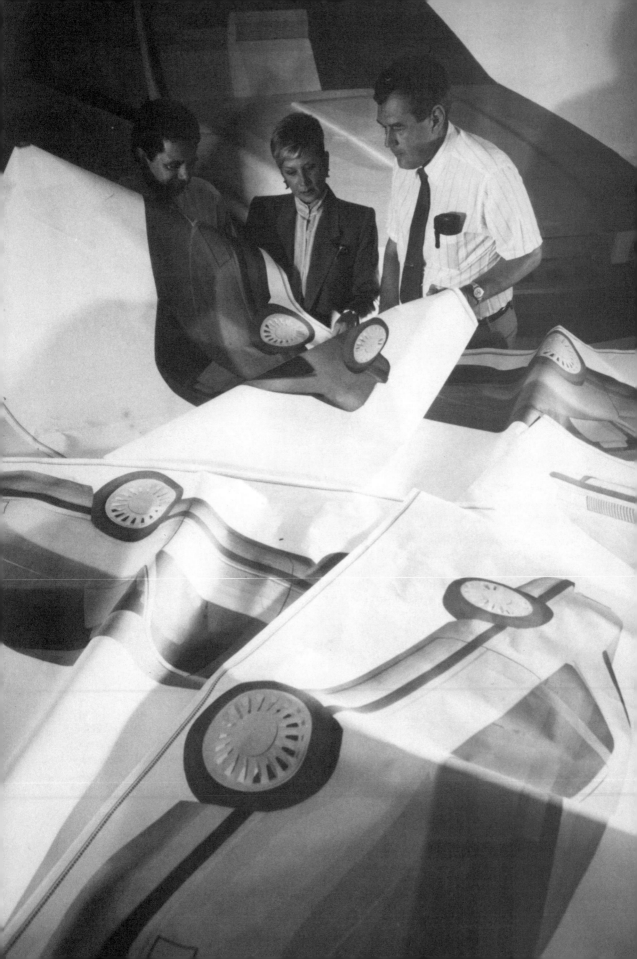

Do you like cars, computers and the excitement of being on the cutting edge of technology? If so, the high demand for people trained in computer-aided design (CAD) could put you in the driver's seat.

A CAD specialist (also called a CAD technician) uses a computer to design a car or the various parts of a car. It's only been in the last ten years that the computer has nearly replaced paper, pencil and drawing board in the auto design business. And though many designs are still drafted on paper, experts estimate that by the year 2000, almost all drafting will be done by computer.

How does the process work? First, engineers decide what they generally want to have in the car part—for example, a camshaft. They pass this information to the CAD designer, who creates a layout of the camshaft on a computer screen. The layout then goes to the CAD detailer who produces a highly specific diagram containing the exact measurements and information needed to build the cam-

1

shaft. CAD detailers can go a step further than those who still work on paper because they can create a three-dimensional model on the computer screen.

At this point the CAD diagram of the camshaft goes back to the engineers. They explore whether anything should be done differently, run some tests and make any changes they feel are needed. The CAD specialist then revises the diagram. When everyone agrees that the part design is what they want, the CAD diagram goes to the machinists, manufacturers and suppliers who will make the camshaft. (In the most technically advanced companies, this information is transferred to the parts makers by computer instead of on a paper blueprint.)

Then a specialist in computer-aided manufacturing (CAM) uses computer systems to actually build the part. The CAM specialist programs, monitors and often repairs the machine or network of machines, including robots, that produce the part.

CAD is more accurate and a lot faster than manually drawing a part, and it creates a permanent record that can be accessed at any time to verify part specifications. Before CAD, designers would draw a part and that would become the master. The model makers would make a model from it; the tool makers would use it to make tools and fixtures to produce the part. But there would be little errors built in all along the way—the slightest mistake would compound itself.

Computer-aided design also speeds up the process of getting a product to market. Today, automobile and parts manufacturers are heavily recruiting potential CAD employees through placement offices at community colleges and vocational training centers. Many companies offer internships and co-op programs to students who are working on their associate degree.

What You Need to Know

- ❏ Computer basics
- ❏ Factory machine operations that produce parts
- ❏ Properties of metals and plastics used in machine parts
- ❏ Basic engineering principles
- ❏ Mathematics (algebra, geometry, trigonometry)
- ❏ Science (chemistry and physics)

Necessary Skills

- ❏ Ability to read blueprints
- ❏ Drafting skills

Do You Have What It Takes?

- ❏ Problem-solving mindset; ability to identify a problem, organize evidence and deduce a probable solution
- ❏ Determination to stay with a problem or project until it is successfully completed
- ❏ Ability to accept criticism without taking it personally
- ❏ Flexibility to make design changes over and over
- ❏ Neatness, accuracy and attention to detail
- ❏ Willingness to spend many hours in front of a computer
- ❏ Ability to communicate well with others—ask questions, take direction and apply it to the situation

Education

Associate degree in auto body design required. Coursework or experience in drafting is helpful. Community college or technical school courses should include descriptive geometry, basic engineering (including hydraulics, pneumatics and electronics), data processing and computer training in computer-aided design software programs.

In the future, major companies may require a four-year degree to obtain an entry-level job.

Getting into the Field

Licenses Required

None

Job
Outlook

Job openings will grow: much faster than average.
The increasing use of CAD/CAM in automobile manufacturing is expected to create a million jobs during the
nineties.

The Ground
Floor

Entry-level job: detailer
Everyone starts off as a detailer, which provides the
on-the-job training needed to move to designer. Some people continue to work as detailers throughout their careers.
Others can move up to designer within a few years.

On-the-Job
Responsi-
bilities

Beginners (CAD detailers)

❏ Meet with designers and engineers on what the car or
  car part will include
❏ Produce three-dimensional diagram of the car part
  design from the designer's layout on the computer
❏ Make changes to the diagram required by the
  designers and engineers

Experienced Designers

❏ Work with engineers to decide what a car or car part
  will include
❏ Produce a computer layout or drawing of how the
  part will generally look
❏ Supervise the work of the detailer
❏ Explore options for the design or other materials that
  could be used
❏ Seek approval from engineers on completed design

When
You'll Work

CAD specialists work a standard 40-hour week. Overtime is often necessary when the deadline for a project is
approaching.

Large companies employing CAD specialists offer two to three weeks of paid vacation and all major holidays off.

Time Off

❑ Pension plans
❑ Stock purchase plans are offered by General Motors Corporation, Ford Motor Company and Chrysler Corporation and some large, publicly held auto parts companies. (In such plans, for every share purchased by the employee, the company buys a share or part of a share for that employee.)
❑ Courses offered at the workplace to keep CAD designers on top of the latest technological developments
❑ Tuition reimbursement for college courses in design, engineering, technology or business management to become more highly skilled or to advance on the job

Perks

❑ All U.S. auto manufacturers
❑ Major auto parts manufacturers

Who's Hiring

Beginners and experienced workers: little or no travel potential

Places You'll Go

CAD specialists work in clean, comfortable and quiet offices. They spend most of their day alone or with one other person at a computer work station.

Surroundings

Because demand for CAD specialists is high, the pay is good. The starting salary for a graduate of a two-year associate program ranges from $18,000 to $24,000. Some companies pay a higher hourly rate for overtime work (if the CAD specialist is an hourly rather than a salaried employee). Top earners in supervisory positions can earn up to $45,000 a year.

Dollars and Cents

Moving Up

CAD specialists advance quickly because of the short-age of skilled workers in this field. Continued education is required to keep up with the fast-changing technology, and companies offer on-the-job training. Many CAD specialists obtain their four-year college degrees in design, engineer-ing or technology while they work so they can become engineers who design and engineer cars and car parts or advance into supervisory positions.

Where the Jobs Are

Jobs are concentrated in the Midwest, particularly Michigan and Ohio, where most automobile and automo-bile parts manufacturing takes place.

School Information

Most high schools and vocational high schools offer drafting courses; some also offer CAD classes. Some for-profit vocational schools, particularly those in the Midwest, offer courses and certificates in CAD.

Community colleges offer CAD courses leading to a two-year associate degree in auto body design. In particu-lar, community colleges in Midwest manufacturing centers have co-op programs (which alternate classroom learning with on-the-job training) with the auto and auto parts companies.

The Male/Female Equation

Traditional drafting and design have been dominated by men, but women are increasingly entering the computer-aided design field. The auto companies are aggressively recruiting women and minorities for positions.

The Bad News

❑ Economic downturns can result in layoffs
❑ Eyestrain from intense computer work
❑ Deadline pressure can be stressful
❑ Jobs confined to a limited geographic area

The Good News

❑ Good pay and benefits
❑ Excellent potential for advancement with additional training and education
❑ Opportunity to work with state-of-the-art technology
❑ Design skills can be transferred to other industries, from architecture to aerospace

◆ **Making Your Decision: What to Consider**

Write for the free brochure, "Automotive Engineering: A Moving Career."

Society of Automotive Engineers
400 Commonwealth Drive
Warrendale, Pennsylvania 15096
412-776-4841

◆ **More Information Please**

WHAT IT'S REALLY LIKE

Cheryl Harris, 21,
detailer, Hawtal Whiting Inc.,
Troy, Michigan
Years in the business: newcomer

What made you decide to go into auto body design?
I have been taking drafting classes since ninth grade. I like
to draw and I took drafting and commercial art as elective
courses because I found I didn't have the creativity and
imagination that commercial art takes. Drafting is good
because you are given guidelines. I work better with
mechanical drawings than free-flowing art.

**Why did you decide to go into the automotive field in
particular?**
My dad works for General Motors. That's the major reason.
Also, in metropolitan Detroit almost all drafting is geared
toward the auto industry.

Did you attend a community college?
Yes. I just graduated from Macomb Community College.
You can elect styling design or auto body design as a spe-
cialty for your degree. My degree is in auto body design.

Did you work while you went to school?
Yes. I was part of the co-op program. The co-op runs in

six-month cycles. You go to school six months and work six months.

How was the co-op program beneficial?

I got more benefit out of the co-op program than I could have gained on my own. I made a lot of contacts. I became friends with the managers, supervisors and other people I worked with in the groups where I was placed. Being in the workplace environment showed me what full-time car company employees do. I also met people from outside design firms who do work for the car companies. So I got a feel for the different environments inside and outside the company.

What was your first co-op job assignment?

My first co-op cycle was supposed to be a manual drafting assignment. I was placed in General Motors' group for passive restraints, which includes seatbelts and airbags. Most of the drafting was done on the computer, but I didn't have any experience on it, so I spent most of my time meeting with people. I did some comparison studies involving different seatbelt systems within the company, and I looked at the European guidelines for seatbelts to make sure our systems followed their restrictions for cars exported to Europe. I met a lot of General Motors' suppliers.

When did you learn to do drafting on the computer?

I had taken a computer graphic systems class taught at General Motors by General Motors people two nights a week for eight hours a week. It was a hands-on experience. I was actually taught by people working on the computer tube, and it was helpful because the computers at school were outdated.

What was your second co-op assignment?

It was also with General Motors, on the team designing the Pontiac Firebird and Chevrolet Camaro. I was placed on the computer in that job because by then I had had computer training. It was great. It helped me get a job after school because I had hands-on experience at General Motors.

Do you feel it is necessary to learn drafting on the drawing board before moving to computers?

Yes. You have to be able to visualize the part you are working on from your board experience to be able to design on the computer. Drafting on the board helps you understand how important the accuracy of your work lines is and helps you to see a two-dimensional object on paper as a three-dimensional object without having the actual part in front of you.

How did you land your first job?

I finished Macomb in May, 1992. I broke in very easily. The people from Hawtal Whiting, which is an outside design firm that does work for the automobile manufacturers, knew my supervisor at General Motors. The people at General Motors helped introduce me to the right people.

What do you do on the job?

I was hired as a detailer. I'm working on the computer tube in engineering on a 1994 mid-size car. Once my skill level can be determined, my supervisor will know how much training I need to advance to layout.

What are you finding the hardest thing about your first job?

Every company has different formats and procedures for doing drawings. The procedure here at Hawtal Whiting is a little different from that of General Motors. But it's not an extreme adjustment.

What is your proudest achievement so far?

With economic times as rough as they are, I'm proud and surprised at how quickly I got a job. I'm ecstatic about that. I called a lot of places to see if they were hiring. This was the first interview I had and I got the job. Macomb gave me great opportunities to meet people, and the co-op program was the best thing that has ever happened to me. I'm convinced the co-op and the people I met through it are the reasons it didn't take me more than a week to get a job.

Do you think computer-aided design is a good field for women?

Yes. So far, I've seen as many women as men in it. Women are getting a fair chance and aren't treated differently.

What is your advice to someone interested in going into computer-aided design?

High school drafting is good but not necessary. If you start drafting in college, you'll have as good a chance as anyone. I'd also advise that you get involved in a co-op program. Take the job you get in the co-op program seriously and do the best that you can. When I was working on my co-op job, I met people from engineering shops who had a lax attitude. They didn't take their jobs seriously and switched from job to job often. Supervisors and managers look highly upon people who work hard and take their work seriously; they want to see good people succeed. My supervisors and managers have been very helpful to me; they've allowed me to use them as references.

Neal Corey, 28, product designer, layout and design section of Powertrain Research, Ford Motor Co., Detroit, Michigan Years in the business: three

Why did you get into computer-aided design?
My long-range goal is to be an engineer, and I knew a design background would make me a better engineer. I also saw it as a good way to get in with a major company and to have my education subsidized, even though it will take me longer to get my degree while I'm working.

What kind of preparation did you have?
In high school I took a college-prep curriculum with a lot of science and math. Then I attended Macomb Community College and took several engineering courses. My associate degree is in auto body design. I was part of the school's co-op program, and I was assigned to Ford Motor. When I was finished in 1989, I was hired by Ford.

Did you take courses in computer-aided design?
Some. I had to take a few classes to get an associate degree, but that barely scratched the surface of what there is to know about the field. CAD is really the technology of the nineties. It is an important skill to have, but I don't think any school should stop teaching drafting. I took three or four drafting classes; you have to learn the principles on the drawing board. It's important to take drafting and de-

scriptive geometry, because the computer is only a tool. If you don't understand how it works on the board, you won't understand it on the tube (the computer monitor.)

Are you a car nut?
I don't particularly love working on cars, but I enjoy the styling aspect of cars. I enjoy knowing how things work, everything from an engine to the door locks. I'm naturally inquisitive. The design end deals more with the artistic aspects of the car rather than the mechanical ones.

What do you currently do?
I work with a small group called Powertrain Research. We produce experimental engines and transmissions and do a lot of tests. We work on concepts more than anything else; we use the newest materials and the latest technology.

How did you start your design work?
I learned on the drafting board. I've only been on the tube about one-and-a-half years. Most of my training was on the job; I took a two-week course for eight hours a day. Ford has decided to move from the drawing board to the computer, and they figure it takes people who are board trained close to two years to work as fast on the computer as they had on the board.

Why is the computer so important?
Its greatest capability is that you can create a three-dimensional model of the part you're working on. You can transfer the data into the computer database and start to manipulate it in different ways. Speed is another advantage. Creating the original design is no faster on the computer than on the drawing board, but the advantage of CAD comes when you make changes. You can quickly update the blueprint. Also, you can share the design information through the computer with other designers who are working on other parts associated with yours.

Are you working on a four-year degree?
Yes. Ford has an in-house program taught by professors from a local university. I'm halfway through the program, and it will take me another three years to finish because I'm only going to school part time. That's all I can handle since I'm working 47 to 50 hours a week.

What kind of personality do you need to be good at computer-aided design?
You have to be very open minded and know how to accept criticism. Not everything you do is perfect. You may design something you think is great, but if it can't be machined, it's not practical. You have to understand what the machines that manufacture the parts are capable of doing and what things they just can't do.

And you've got to have patience—there are a lot of changes before parts are finalized. Every large part that you make will get changed somewhere along the line. You have to accept and understand that or you'll be extremely frustrated.

What do you like most about your job?
It's challenging. If there's a problem, you have to figure out a way to solve it. You can come up with great ideas, but they may not always be feasible.

What do you like least?
There's not much I don't like. But working in a small research group within a large corporation can be frustrating because so few of our ideas ever reach fruition. We've been working on a project now for two years and it may go no farther than our group.

What is your advice to someone who's considering a career in this field?
While you're in school, work not only on your design skills but also on your science and math. You have to understand the concepts and principles you're working with because the technology is changing so quickly. You also must develop good communication skills and good people skills so you can get your ideas across in a manner that's easy for everyone to understand.

What has been your proudest achievement?
I'm proud of my ability to work well with everybody. We're a team here. It's nice to get a raise and a good performance review, but you can't do it at the cost of everybody else. There are 300,000 people working for Ford Motor Company and we have to work toward a common goal.

John Viviano, 41,
manager of computer graphics,
VEHMA International of America, Inc.,
a division of Magna International,
Detroit, Michigan
Years in the business: 23

How did you get into automobile design?
I always liked to work with my hands and create things.
And I liked drawing. When I was in high school, I took a
basic drafting course and then decided to get into graphics.

Were you interested in automobiles, specifically?
When I was in high school, I didn't know what field I
wanted to go into because I wasn't familiar with what was
out there. I didn't even know how a car was built. But as a
kid I do remember going to the dealership in September to
see the new cars on the lot and to see what kinds of engines
they had in them.

**You went to college before computer-aided design was
being used, right?**
That's right. I learned manual drawing when I went to Macomb Community College. I have an associate degree in
auto body design and tool fixture, and I have a certificate
in die design. I didn't start CAD until 1983.

Did you take part in a co-op program?
Yes. I worked in a couple of engineering shops, firms that
do design and engineering for the auto companies. My
second co-op job was with Ford Motor Company. After I
graduated, they offered me a job full time.

What did you do in your first job at Ford?
I started as a detailer in the body engineering division. I
worked with designers, pulling details off the design plates.
Back then the designers worked on aluminum plates with
white enamel paint, and they inscribed the gold into the
white paint. The detailer would take a piece of mylar and
trace the part he was detailing right off the surface plate to
create a detailed drawing to Ford's standard. With today's
computer technology, that's obsolete.

Did you continue in that department?

I climbed the ladder in that department to senior designer. Then in 1972 I was laid off. That's the only time I've ever been laid off in my career. A month later I was picked up by Ford's truck engineering division.

When did you leave Ford?

I left in 1986 and started working for a company called Modern Engineering. It's a service firm that does engineering and manufacturing for the Big Three auto companies: General Motors, Ford and Chrysler. I was a design leader for their CAD/CAM activities.

How did you learn CAD?

I had been trained on Ford's CAD/CAM software and hardware systems before I left. I took a 12-week course for 20 hours a week. I worked full time on CAD at Ford for three years. Then I was full time on CAD at Modern for two years and started the CAD department for them. I left Modern in 1988 and went to Magna International, the mother company of VEHMA. It also does engineering and manufacturing for the auto companies.

Describe your current job.

I am responsible for computer graphic software and hardware. In a normal day I make sure the work load is on schedule, I assign work, I work with customers. I do the scheduling of work on a weekly and monthly basis. I'm responsible for the performance of seven CAD specialists. I also have shipping and receiving, and I'm responsible for coordinate measurement machine inspection for prototype parts.

What kinds of qualities do you look for in CAD specialists you hire?

The job candidate must be trained properly for the specific system we use here. Unfortunately, a lot of CAD specialists only know the particular system they learned on.

Is an associate degree required to get your foot in the door?

Let's put it this way: Job experience is more important than a degree. But a degree—either an associate degree or a bachelor's degree—is useful. Experience is very important because the person has to be productive. The auto industry is changing and the pressure is on to cut the time to do a project and to increase productivity.

What do you like about your job?

I like talking with the customers—people from the auto companies that we do work for. I like getting their input, passing it along to the CAD specialists and seeing it incorporated into the design. That's what teamwork is all about.

What is the biggest challenge in your field?

The constantly changing technology. It's moving so fast that sometimes it's off on the next step before the previous step has been proven effective. The limited amount of time and money we have to get a project done can also make the job difficult.

What achievement are you particularly proud of?

I'm proud that I helped create engine packaging, fuel tanks or sheet metal that's still on cars and trucks I see going down the street to this day. I can say I designed that fuel tank on that truck or I put that motor in that car. It makes me proud to see my work on the road.

What advice would you give someone interested in the CAD field?

I'd advise the person to get full training from a certified training center for the CAD software and hardware used in the auto industry. I also suggest they be trained in drafting and design on the drawing board. I strongly believe that to use the CAD system efficiently, you have to know how to draw on the board.

The days of the stereotypical fast-talking car salesman in the gaudy plaid jacket are gone. Today's car salesperson is a professional who not only sells the car but is also an ambassador for the dealership. If you're a fan of automobile shows, find yourself talking about the latest models on the market and love to test drive anyone's new car, a job in automobile sales may be for you.

P revious sales experience is a big plus; salespeople from other businesses often move into car sales because they can make more money through bigger commissions. If you have no previous sales experience, a common first job in automobile sales is on a dealership's used car lot.

Since the salesperson is the first contact the customer has with the dealership, he or she has the best shot at mak-

19

ing a good impression. Customer satisfaction is the name of the game in the automobile business today, so you need to pay attention to what customers are saying and asking for.

To be successful, it helps to be outgoing and persuasive. You must be willing to work long hours and make cold telephone calls to people you don't know who might turn out to be prospective customers. You must be good at building relationships with customers so they'll come back to buy their next vehicle from you or refer people they know to you. If you're good, you can make big bucks... and you can drive home a different model every night if you want to.

What You Need to Know

- ❑ What the various parts of a car (or van, truck or other vehicle) are called and how they work
- ❑ A general understanding of how the automotive industry works and your specific competition
- ❑ History of the cars/models you're selling
- ❑ Car finance options so you can make recommendations to customers

Necessary Skills

- ❑ Sales experience helpful
- ❑ Basic math skills (to figure out monthly payments on a car at various interest rates and loan periods)
- ❑ Familiarity with a calculator
- ❑ Good driving skills

Do You Have What It Takes?

- ❑ A friendly, outgoing personality
- ❑ A well-groomed appearance
- ❑ The power of persuasion
- ❑ Ability to maintain your cool when customers become difficult or rude
- ❑ A positive attitude that allows you to shrug off rejection
- ❑ Strong legs—you'll often be on your feet most of the day
- ❑ A good telephone manner
- ❑ Discipline to organize your workday
- ❑ Motivation to initiate sales calls
- ❑ Ability to negotiate and make compromises so that you can close a deal

Education

A high school diploma is required. Additional education may give you an edge over other job candidates. Courses in sales and business management are helpful.

Licenses Required

Driver's license. Clean driving record.

Job Outlook

Although the number of car dealerships will continue to decrease because bigger dealers are buying up smaller dealers and merging the operations, openings in car and truck sales will remain numerous because of the high employee turnover rate. Jobs are very competitive at dealerships that carry top-priced cars and the models in highest demand.

The Ground Floor

Entry-level job: salesperson

Large dealerships have a number of entry-level positions that can help prepare you for a job in new vehicle sales. They include:

❑ Used car salesperson
❑ Service advisor (writes up repair orders and sells the services a dealership provides)
❑ Parts and accessories salesperson

On-the-Job Responsibilities

Beginners

❑ Meet prospective customers at the door or on the car lot
❑ Explain the technical aspects of the car and its advantages over the competition
❑ Discuss optional features available on the car
❑ Go over warranties, financing arrangements and insurance programs sold by the dealership
❑ Take customers on demonstration rides and drives
❑ Negotiate the final sale
❑ Maintain a list of prospects and regularly contact customers
❑ Meet sales goals (set by the sales manager on a weekly, monthly or quarterly basis)

Experienced Employees (sales managers)

Do all of the above, plus
❏ Manage and motivate the sales force
❏ Provide training to sales staff
❏ Assist with negotiations
❏ Approve the final deals
❏ Hire and fire salespeople

Experienced Employees (fleet managers)

❏ Deal with business customers who buy several vehicles at a time

When You'll Work

Get ready for long hours—often 60 to 80 hours a week. Salespeople usually work Monday through Friday and one day of the weekend, including some evenings. Some dealerships give salespeople a day off during the week.

The busiest times of the year are the fall, when new vehicles are introduced by the manufacturers, and spring, when the winter doldrums are over and better weather lures customers into the showroom. But whether customer traffic is slow or busy, the showroom must be staffed constantly.

Time Off

Vacations are usually scheduled for the winter months, when sales are traditionally low. Beyond that, only major holidays—Thanksgiving, Christmas and Easter—are paid days off.

Perks

❏ Manufacturer discounts (about five percent) on vehicles sold by the dealership
❏ Overnight use of new dealership vehicles

Who's Hiring

❏ Dealerships employ nearly a million people. In 1992 it is estimated that there are 24,000 new car dealerships in the U.S.
❏ Used car dealers

Places You'll Go

Beginners and experienced salespeople: little travel potential.

The exception is a trip for an occasional training seminar offered in locations that can be reached by car.

Surroundings

The car salesperson works in a clean, well-lighted environment with large windows looking out on the dealership property, which is usually located on a busy avenue or in the central business district. He or she is likely to have a desk or a small office off the main showroom. The salesperson also must go outdoors frequently to show vehicles that are parked there.

Dollars and Cents

The average car salesperson earns $19,900 a year. However, a well-established salesperson who attracts many return customers can earn as much as $100,000 a year. Most salespeople are paid on commission, that is, a percentage or a flat fee for each vehicle they sell. A percentage commission is usually about 25 percent of the dealer's gross profit on the sale, so the better the deal the salesperson negotiates, the bigger his or her profit. A large luxury car has a higher profit margin than a small, low-priced car. A very popular model has a higher profit margin than does a slow seller.

The flat fees that some dealerships pay usually range from $50 to $75 for each car or truck sold. Some dealerships provide a minimal base salary in addition to commissions. Others pay a base salary if the commissions for the week are lower than the salary. Some provide a minimal base salary to new salespeople and gradually decrease and eliminate the salary once the salesperson is established and is regularly earning commissions.

Moving Up

The more vehicles you sell, the better your chances of moving into a manager's spot, that is, supervising the work of other salespeople. In addition to being a super salesperson, you must also be able to work well with all kinds of people, know how to delegate work, and have the skill and enthusiasm to motivate others.

Earning your certification from the National Automobile Dealers Association, the industry trade group for new car dealerships, is useful for advancement or for making a move to another dealership. Taking classes on your own in business management, accounting or sales training can also help you improve your skills and shows your bosses that you're serious about getting ahead.

The next step above sales manager is general manager, the person who oversees all of the operations of the dealership. Some dealers reward good managers by making them a part owner in a dealership or helping them get established in their own dealership. Most dealers begin their careers in sales.

Dealerships are located in every city and town across the U.S., so sales jobs are widely available. More dealerships are located in large cities, and, not surprisingly, they often employ more people and handle a bigger volume of sales than those in outlying areas.

◆ **Where the Jobs Are**

All automobile manufacturers provide training to salespeople at their dealerships. That training can involve anything from watching a videotape at the dealership to a week-long seminar at an out-of-town location to learn all about the manufacturer's new product line.

◆ **Training**

The National Automobile Dealers Association's training and certification program for salespeople consists of 12 hours of self-study and 8 hours of classroom instruction on ethical and legal practices, selling techniques, consumer psychology, customer loyalty and sales manager training.

Men far outnumber women in automobile sales, but women are increasingly entering the field. They often have an advantage in selling to other women, who buy half the cars sold today.

◆ **The Male/Female Equation**

Making Your Decision: What to Consider

The Bad News

- ❏ Usually no base salary or pension
- ❏ Long hours
- ❏ Earnings tied to ups and downs in the economy
- ❏ Stiff competition for customers
- ❏ Stereotype of being smooth talking and deceptive

The Good News

- ❏ Potential for high earnings
- ❏ Earnings relate to sales ability
- ❏ Good chance of advancing to management spot
- ❏ Potential to own your own dealership
- ❏ Personal car discounts

More Information Please

Write for the free brochure, "Automotive Careers," which describes various careers in the retail end of the automobile industry.

National Automobile Dealers Association
8400 Westpark Drive
McLean, Virginia 22102
703-827-7407

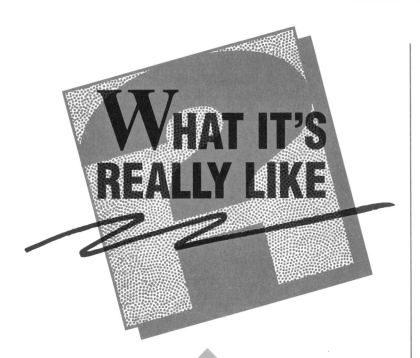

WHAT IT'S REALLY LIKE

Linda Finch, 45,
salesperson, Spirit Nissan Jeep/Eagle Inc.,
Sylacauga, Alabama
Years in the business: one

How did you break into the field?
I had worked at a number of jobs and couldn't find anything that really satisfied me. I was a telephone operator
and a police officer. I worked in the textile industry. Then
I took some abilities tests and the results showed that I'd be
good in sales, so I started selling hearing aids door to door.
Then I went into insurance sales. A friend was selling automobiles and talked me into taking her job at the dealership
when she left.

When I started, I stood at the door to catch every person
who walked on the lot because every looker is a potential
buyer. It's a great way to meet people. Then I started using
the phone—I had a lot of contacts from the insurance business who became a source of business for this job. I also
did cold calls—telephoning strangers to find out if they
might be in the market for a new car. It's hard, but I was
used to doing it from the insurance business.

What kind of preparation did you have for sales?
I have a high school education. I received a lot of formal training from the insurance company where I worked, including a two-week program where I learned memory aids and worked on developing a positive mental attitude and self-confidence.

What was the hardest part of your job at first?
Learning about cars. All I knew was that you drive them and change the oil. Now I can tell you everything about the cars in this dealership.

How long did it take you to get established?
You never get established in the car business. If you want to remain successful, you can't think of yourself as being established. There's always something new to learn. You always have to dig for new prospects. I like to keep ten new prospects on a list, and if I sell to one or lose one, I find another. You have to have a goal in front of you because you don't get paid unless you sell.

What do you like most about your work?
The people. You get to meet lots of different people.

What do you like least?
The long hours. I put in ten good hours a day, five days a week.

What have been your proudest achievements?
I really enjoy watching customers drive away in a brand-new vehicle that they totally love, knowing that I put them in that vehicle.

What advice would you give to someone who is thinking of going into this field?
Go into it if you've got a positive attitude. It can be hard, really hard, because rejection is an everyday thing. And when you start out, don't make it your household's major source of income. You need time to build a clientele. If it's your sole source of income, you'll put too much pressure on yourself to succeed.

Eric McHenry, 28,
salesperson, Sewell Village Cadillac,
Dallas, Texas
Years in the business: seven

How did you break into car sales?
I started in the accounting department of the dealership.
Then I moved into the service department and drove the
courtesy car (taking people from the dealership to their
homes or jobs while their cars were being repaired). My
next job was sales assistant, selling used cars, and then
finally, selling new cars. I've been doing that for four years
now.

What kind of preparation did you have?
I have a high school diploma. I went to community college
for two years, but I don't have a degree. I got my sales
training at the dealership.

**What was the hardest aspect of working in this field
during your first few years?**
The rejection is tough. I was in sales jobs before, but they
weren't commission based. Any time someone says ''No,''
that's money lost.

How long did it take you to get established?
When I was selling used cars, every month was a better
month. I made more money and I learned more than the
month before. The lessons I learned were almost as important as the money I earned. When I went into new car sales,
I felt I had arrived.

What do you like most about your work?
I love the cars, but cars are still just cars. I love working
with customers. You meet the most interesting people from
all walks of life. None are the same, and they all have different things that make them click.

What do you like least?
The hours. I work from 8 A.M. to 8 P.M. Monday through
Friday and 8 A.M. to 6 P.M. Saturday. That's 70 hours a
week.

What has been your proudest achievement?

Going to Europe. The dealership I work for has an achieve-ment club called Team 20. If you sell 20 cars in any two months of the year, you can build up enough points for trips. I went to Europe for 14 days. It was a dream I had long ago, but I never thought it would really happen. I came from a neighborhood that wasn't affluent and I don't have a college degree. To make this kind of money and be awarded with a trip to Europe is a dream come true.

J.R. Arnold, 52,
salesperson, Hoffman Chevrolet-Jeep/Eagle, Hagerstown, Maryland
Years in the Business: 13

How did you break into car sales?

I spent 21 years in the Navy and interviewed for several different jobs when I got out. I was 39 at the time, so it wasn't easy convincing people to hire me. I finally got a job selling used cars at another dealership. After one year, they offered me a guaranteed salary to work in the truck department. I took it.

After that I worked in sales and management at other deal-erships, but then I came back here in sales. Now I also oversee the truck department, although my title is salesperson.

What kind of preparation did you have?

When I was six years old, my mother encouraged me to play the violin for audiences. That gave me confidence to perform in front of people. In the Navy I worked in meteo-rology and had to brief pilots. You have to be knowledge-able and make them feel confident with your flight fore-casts. I also had to give weather briefings in front of audiences. Some of the other guys offered to pay me to do their briefings because they were afraid to talk to groups and I wasn't.

What was the hardest aspect of working in car sales in your early years?

I tried to make every customer my personal friend. I found

out that was a mistake. Customers are out to get the best deal and will stop at nothing to do it. You can tell them their interest rate will be between nine and ten percent, and then they'll go tell the finance person you promised them eight!

How long did it take until you were making a good salary?
My first week on the job I made $750. In the first year I made $16,000. I could see it going up year after year.

What do you like most about your work?
I like the vehicles. I've always been a car nut.

What do you like least?
Haggling with customers to make a deal. The customer has been brainwashed into thinking that dealerships in the larger cities can make better deals. They don't look at the big picture; a smaller dealership like ours can provide more personal service.

What has been your proudest achievement?
I've won a lot of awards, but I'm most proud of the monetary rewards and what they have allowed me to give my family. I've made as much as $50,000 a year. That's pretty good for someone with only a high school education.

What advice would you give someone who is thinking of being a car salesperson?
If you're not willing to work 12 hours a day Monday through Friday and weekends, you'd better stay out of the business. If you're the kind of person who takes rejection personally and doesn't bother asking again if someone tells you "No," then selling cars is not for you.

If you spend every Saturday afternoon tinkering under the hood of your car, if you'd rather figure out for yourself what's making that knocking sound instead of paying someone else to, and if your friends tell you they don't trust anyone else but you to fix their car—then maybe you should consider turning your pastime into a profession.

As long as there are cars there will be a need for people to fix them. But gone are the days of the "grease monkey" wielding a wrench. Automotive technology has become so complex that even backyard mechanics must take their cars to specialists for repairs and maintenance.

It is estimated that there are 771,000 automotive mechanics (or service technicians, as they're sometimes called) in the United States, but there's plenty of room for more. There will be even greater demand for mechanics in

33

the future because more cars will be on the road and more people will try to keep their cars for longer periods of time.

Most mechanics work for automobile dealers, independent automotive repair shops and gasoline service stations. With the increased complexity of the automobile, a mechanic frequently specializes in areas such as automatic transmissions, tune-ups, front ends, brakes or air conditioning and heaters. Mechanics can also devote their practice to vehicles other than cars, such as diesel engine trucks, buses, motorcycles or boats.

With thousands of car accidents guaranteeing a steady supply of work, some mechanics go into automotive body repair. They can specialize in this field as well, focusing on frame straightening or door and fender repairing, glass installation, fiberglass repair or painting. Some workers especially enjoy customizing and refurbishing antique or "collector" cars.

You can break into the business with only a high school diploma and some body shop or automotive courses. Increasingly, however, employers are looking for formal training in automotive mechanics beyond high school.

If you like to hear the smooth purr of a well-tuned engine—and don't mind a little dirt under your fingernails—read on.

What You Need to Know

- ❏ Extensive knowledge of car mechanics and electronics
- ❏ Knowledge of computers (for diagnosing engine and electrical problems)
- ❏ Familiarity with tools
- ❏ Basic math (to calculate car repair costs)

Necessary Skills

- ❏ Mechanical know-how (ability to take apart objects with moving parts and put them back together again)
- ❏ Analytical skills (figuring out where a problem exists and how to solve it)
- ❏ Good communication skills (ability to talk to, ask questions of and explain technical things to customers and service advisors)

Do You Have What It Takes?

- ❏ Manual dexterity (ability to handle small tools and parts easily)
- ❏ Patience to deal with difficult or demanding customers
- ❏ Willingness to get your hands dirty
- ❏ Strong powers of observation (to eyeball which parts may not be functioning correctly)
- ❏ Ability to stick with a repair until it's finished

Education

A high school diploma is usually required. Auto repair courses (offered in vocational/technical schools or community colleges) are advisable.

Licenses Required

Driver's license

Job Outlook

◆   **Job openings will grow:** much faster than average.
  The supply of auto mechanics is far below the demand, so there are plenty of opportunities for mechanics who complete training programs.

The Ground Floor

◆   The entry-level job for an auto mechanic varies, depending on the training the person already has. The following jobs can lead to that of automotive mechanic:

❑ Porter or car washer. Helps prepare new and used cars for delivery and performs odd jobs within a dealership.

❑ Gasoline service attendant. Pumps gas, checks oil, performs minor repairs.

❑ Trainee, apprentice or helper. Works with an experienced mechanic, shop foreman or service manager. May lubricate cars and do light repairs.

On-the-Job Responsi- bilities

◆ *Beginners*

❑ Wash cars
❑ Drive cars in and out of the service bay
❑ Lubricate cars
❑ Do light repairs and routine maintenance

Experienced Mechanics

❑ Talk with customer or service advisor about what is wrong with a car and what needs to be done
❑ Diagnose a problem based on information received from customer or service advisor, results of tests with diagnostic machines or a test drive of the car
❑ Provide a cost estimate of parts and labor needed to fix the problem
❑ Make adjustments and repairs, replacing parts that are broken or damaged
❑ Inspect car systems during routine maintenance checks

Automotive mechanics generally work 40 to 48 hours a week. Overtime may be necessary if there is a lot of work and/or a shortage of mechanics. Self-employed mechanics work longer hours. Mechanics usually start early in the morning, when most people drop off their cars for repair. Some dealerships and repair shops are open evenings and Saturdays. Mechanics usually have Sundays off.

◆ **When You'll Work**

All major holidays and one to four weeks of vacation a year, depending on the employer's policy.

◆ **Time Off**

❏ Retirement plan (about 20 percent of new car dealers offer a retirement plan)
❏ Manufacturer discounts on vehicles. Amount varies by manufacturer and the particular vehicle but is usually about five percent.
❏ Discounts on auto parts

◆ **Perks**

❏ Dealerships for new and/or used cars
❏ Independent repair shops
❏ Gasoline service stations
❏ Automotive service facilities at department, automotive and home-supply stores
❏ Companies maintaining fleets of cars and trucks, including rental car, taxicab and automobile leasing companies
❏ Federal, state and local governments
❏ Motor vehicle manufacturers
❏ Parts manufacturers and wholesalers

◆ **Who's Hiring**

Beginners or experienced mechanics: little or no travel potential.
Employers sometimes send mechanics to factory training centers across the U.S. to learn to repair new models or to receive training in the repair of special components.

◆ **Places You'll Go**

Surroundings

Mechanics work indoors most of the time, occasionally going out to test-drive a repaired car or to pick up or deliver a car. Modern repair shops are well lit and well ventilated, but the environment is often noisy because of all the equipment.

The work can be dirty and greasy, and minor cuts, burns and bruises are common. The mechanic may have to get into awkward physical positions to repair hard-to-reach parts of the car. Some heavy lifting of parts and tools may be required. The mechanic must work with hazardous substances, including oil and transmission fluids, solvents and paints. Some jobs require the use of special equipment, such as respirators for painting and goggles for welding.

Dollars and Cents

The average mechanic at a new car dealership earns nearly $24,000 a year. The average starting salary is about $17,000, and the average top-earner makes about $33,000.

Highly skilled auto mechanics employed at automobile dealerships in 18 metropolitan areas earn an average of $18 an hour. Less skilled mechanics doing routine service and minor repairs earn an average of $12 an hour. Semiskilled mechanics earn $9 an hour. If you work overtime, you'll often earn a higher hourly wage. Mechanics with certification or those with a specialty make more per hour.

Some dealerships and repair shops pay on a commission or profit-sharing basis so that the mechanic's weekly earnings depend on the amount of work completed. Some pay the mechanic a percentage of what the customer is charged. Under this arrangement, employers often guarantee a minimum weekly salary if commissions fall below the salary level.

Moving Up

Auto mechanics have tremendous opportunities for advancement. Experienced mechanics with strong leadership skills and good business sense can advance to shop supervisor or foreman, service manager, body shop manager and then general manager. A general manager can become owner or co-owner of a dealership. A good mechanic always has the opportunity to open his own business. About one in five automotive mechanics today is self-employed.

Jobs are available across the U.S. Every town has a dealership or repair shop that needs mechanics.

◆ **Where the Jobs Are**

The best jobs go to automotive mechanics who complete a formal training program after high school. The post-high school training programs vary in length, from six months to two years, and usually include a combination of classroom instruction and hands-on experience. Training programs are offered by community colleges as well as public and private vocational and technical schools.

◆ **School Information**

Automobile manufacturers and their participating dealers sponsor two-year associate degree programs in auto mechanics at about 80 community colleges across the country. The manufacturers provide equipment and cars for students to use for practice.

Voluntary certification—which can mean better pay and better jobs for a mechanic—is offered by the National Institute for Automotive Service Excellence. The certification is recognized as a standard of achievement for mechanics, body repairers and painters. Mechanics are certified in one or more of eight different service areas, including electrical systems and engine repair. A master automotive mechanic is certified in all eight areas. Mechanics are retested every five years.

The vast majority of mechanics are men.

◆ **The Male/Female Equation**

Making Your Decision: What to Consider

The Bad News

- ❏ "Grease monkey" job stereotype
- ❏ Skills require constant updating because of new techniques and equipment
- ❏ Regular exposure to hazardous chemicals
- ❏ Work can be dirty and greasy

The Good News

- ❏ Easy to break in
- ❏ Job security; career not likely to be affected by bad economy
- ❏ Satisfaction of seeing the results of your work
- ❏ Plentiful job opportunities

More Information Please

For information on training and certification for auto mechanics and auto body repairers or to obtain a list of certified schools for automotive mechanics and body repairers in your area, write:

National Institute for Automotive Service Excellence
13505 Dulles Technology Drive
Herndon, Virginia 22071-3415

To obtain the brochure "Automotive Careers," write:

National Automobile Dealers Association
8400 Westpark Drive
McLean, Virginia 22102

WHAT IT'S REALLY LIKE

Timothy Clark, 20,
automotive technician, Schaller Oldsmobile,
New Britain, Connecticut
Years in the business: One and a half

How did you get into automotive mechanics?
When I was 12 years old I started working on cars with my
father. Then I took an auto shop class in high school. When
I graduated I went to Denver Automotive and Diesel College in Colorado for 15 months and received an associate
degree.

What courses were included in your program?
I took a management course and a success skills class that
taught me how to manage money. In engines class we completely disassembled and reassembled an engine. Each class
was six weeks long, four hours a day. I also have ASE cer-
tifications in engine repair, air conditioning and standard
transmissions.

How did you get your first job?
The job placement office at my school told me where to
look in my area. I took a resume to one dealership, interviewed two days later and was hired. It was my first
interview.

41

I started right away doing simple things like rotating tires, changing oil and transmission fluid and correcting simple wiring problems like windows that wouldn't go up and down. After about three weeks I started doing heavy engine work—working on intake manifolds, brakes and suspensions, transmissions—and diagnostic work on computers.

What was the hardest part of the job at the beginning?
I was scared I'd mess up an expensive piece of equipment.

What do you like best about your job as a technician?
I love the auto industry, and I love to work on cars. When a car comes in blowing smoke or leaking oil, I fix it as best I can. It leaves the shop in perfect condition. That gives me satisfaction.

What do you like least?
I don't like how dirty you can get on certain jobs. I don't like the grease and cuts. You can make very good money if the work is there, but the pay on certain jobs isn't good.

How did you become an automotive mechanics teacher?
My high school teacher saw my name in the paper when I graduated with a 3.8 grade point average from the community college. He asked if I would teach an adult education course two nights a week. I teach basic maintenance, like oil changes and tire rotation. I teach the people in my class how to relate to the people at the dealership or repair shop.

Do you want to be a technician your whole career?
I want to go into computer diagnostics. I'm trying to go back to school for an electronics degree. I want to get out of the mechanical and into the technical area. I like the challenge of diagnosing a problem—like an engine that skips intermittently at 50 miles per hour. You have to figure out the problem by what information goes into and comes out of the computer.

What is your proudest achievement?
I completely restored a 1980 Chevrolet Camaro from the ground up. It was falling apart. My father gave it to me when I was 15 years old so that I could get to know more about cars. I started off with basic repairs and drove it to Colorado for school, where I rebuilt the engine. Then I took the car home and restored it piece by piece. I took the

body off the frame, redesigned the wiring the way I thought it should be, and rebuilt it. It took a year and a half of working on it a little every day of the week. This year I hope to enter it in car shows.

What advice would you give to someone thinking about going into auto mechanics?
Get as much experience and background as you can before going to a trade school. Then when you get to school, you'll be ahead of the other students and can work on more challenging projects.

John Wolfe, 32,
executive manager, Hoosier Oldsmobile-Cadillac-Honda, Richmond, Indiana
Years in the business: 13

How did you become interested in the auto industry?
When I was 14, I started working in the car wash business in Anderson, Indiana. I fell in love with cars. When I was 15, I was managing the car wash. Then at 16 I started to work in the service station part of the car wash so I could learn the mechanical side of things. A year later I was managing the service station.

What kind of preparation did you have?
After I graduated from high school, I went to Lincoln Technical Institute in Indianapolis to take automotive and diesel technology. I took a double load so I could get through in 9 months instead of 18. After I graduated, I went back to the service station for a few months, but I was bored. I was tired of changing oil and transmission fluid. I wanted to get into electronics. I went to Muncie (Indiana) to all of the dealerships and picked the one with the best appearance; I applied there for a job. I got a job changing oil. Then I moved on to more complicated repairs.

How did you get into management?
Eight months after I started, I was a shop foreman. Then I became a service advisor. Then the service manager left. I

was only 22, so the dealer decided to look elsewhere to find a new service manager, but he gave me the opportunity to be the interim service manager until he found one.

About a year and a half after that, I became the fixed operations director. That meant I was in charge of service, parts and the body shop. In a dealership there is the sales side, which is called the variable part of the business, and the service side, which is the fixed side. I did that until June 1990, when I became fixed operations director of two dealerships, including the one I started with. The same management group then bought a third store. I was instrumental in the startup, the hiring and setting up for the new store. Once it was opened, I was in charge of fixed operations for all three dealerships. Not long after, the management group sold one of the dealerships, so I was involved in laying off people, closing out accounting, taking final inventory and then selling to the purchaser.

How did you become the executive manager?
I continued as fixed operations director for the remaining two dealerships. In June 1991 I went to General Motors Dealer Development school. I'd been waiting to go for three years but kept canceling because of the business activity. While I was attending the school, I got a call from a friend who was running one of the dealerships. He told me the dealer I worked for had purchased another dealership and wanted to put me in charge.

I was scared. I didn't have any sales experience. I went through all kinds of emotions—I felt excited, panicky and physically ill. I came back from school on a Friday, and we had to open on Monday morning. I spent Saturday interviewing employees and preparing for the opening.

How did you learn the sales side of the business?
It was trial by fire. I'm learning it now while I'm in a management position.

How did your background as a mechanic help you?
It taught me an appreciation for the technology in our business and the experience level needed to repair vehicles. I remember I wanted to take auto shop in high school, and my guidance counselor said I was "too smart." Auto shop wasn't a serious course. That's the problem now—we're

guiding talented people away from our industry.

What do you like most about your work?
The challenge. It's never the same, not only from day to day but from hour to hour. This week we're restructuring the store in terms of the number and types of jobs we have. I'm aggressive in my ideas and concepts. I'll try anything once, particularly in fixed operations.

What do you like least?
The hours. I'm working 70 hours a week to get the dealership established. Even once you get established, a 50- to 60-hour week is normal.

Do you feel established now?
No. The day you feel established is the day you start going backward, and then it's time to get out of the business. Otherwise, everybody will start passing you up. The auto business is going through an evolution. It will never be the same as it was five years ago. It takes day-to-day adjustments to stay in the business.

I do feel successful, though. When I became a fixed operations director at the first dealership, I was only 24 years old. I felt I had accomplished quite a bit very quickly. In fact, I would feel successful if I were still doing that today. It was such an all-encompassing job. The service side of the business is very complicated. It's not just fixing cars any more.

What is your proudest achievement?
What I was able to accomplish by the age of 31, when I got to run this dealership.

What advice would you give to someone who wants to be a service technician?
Be patient with yourself and your employer. If you go to a technical school, don't expect that you'll come out and make a ton of money right away. There's a lot of hands-on experience necessary to become proficient.

Successful people often take two steps forward and one step back before moving ahead again because they are willing to take risks and fail. You have to have the self-confidence and ego to fail and then say, "Let's try again."

Daniel Bailey, 40,
owner, Daniel's,
Birmingham, Alabama
Years in the business: 20

When did you become interested in cars?
Ever since I can remember, I've been a car enthusiast and a "horse trader." I've had a lifelong thirst for taking things apart and putting them back together. When I was two, I took apart a clock without damaging it. When I was three or four, I asked for a real wrench, not a play wrench. Today I consider myself an automotive rebuilder or restorer. I take off every bolt from front to back of a car and put it back together.

One of my goals in life was to own a Ferrari—not to drive or ride in one briefly, but to own it. On October 29, 1984, I bought my first Ferrari, a 1977 308GTB, which belonged to Nigel Olsen, Elton John's drummer.

What kind of preparation did you have for being a mechanic?
I wanted to become a certified mechanic, so I took the two-year course at North Georgia Tech in Clarksville, Georgia. After I was in school for a year, I was drafted and joined the U.S. Army Reserves. I became a mechanic, servicing everything from Jeeps to earth movers. After six months of active training I went back to school and finished.
At that point I wanted to open my own automobile shop, but I knew I needed to learn more about the ins and outs of running a successful shop. At age 19 I became the youngest Snap-on Tools franchise owner.

When did you open your present shop?
In 1976 I found a building to rent and I opened under the name of Classic Car Care, Inc. The name was chosen with great consideration. Classic, meaning the finest or best. Car, meaning automobile. Care, meaning to pay close attention to or to feel concern for. I started my business because I felt like no one cared about giving cars the attention and service that I felt they deserved. On the tenth anniversary of the shop we had to change the name because a chain of car washes opened here with the same name.

When did you feel successful?

In the seventh year of the business. One evening after working late, I sat on my workbench and looked around my shop. All of a sudden I understood what success was for me. I was making a living doing what I loved the most, working on the finest cars in the world. Since then I've had that feeling of success several times.

What do you like most about your work?

I like to take a "dog" and turn it into something neat. I love to take the little car that was a piece of junk and when it's finished have people ask, "Where's that junk car you had here?"

What do you like least?

The customers. They are caught up in the world of time, always in a hurry for the work to be done. You can't rush art.

What is your proudest achievement?

I've worked on all kinds of cars—Aston Martin, Mercedes, Porsche, Ferrari, Rolls Royce. My proudest achievement was when I brought back to life a 1957 Mercedes-Benz 300SL Roadster. It was towed to us. I completely disassembled and rebuilt it, interior and exterior. It took six months to rebuild the engine alone.

What advice do you have for someone thinking about becoming an automobile mechanic?

You have to realize that you are in a state of constant learning. After you've been in it for a while, you realize that if you think you know it all, you will learn nothing more.

As the saying goes, "Accidents will happen." And when they do, an auto claims representative is never far behind. It's his or her job to decide how much money people are entitled to receive on their insurance claim after a car accident. Although the procedure is always the same, the people and events never are.

A claim may be filed after a simple fender bender, or there may be extensive damage and serious injury, even death. And sometimes an accident may not be an accident at all but a fraud.

There is a variety of jobs within the field. As an auto claims representative or adjuster, you determine whether losses or damages from an auto accident are covered by the insurance company's policies. If they are, you estimate the costs of the repair or replacement. Then you work out a settlement that is fair to both the policyholder and the insurance company. You negotiate with the auto body repair shop or dealership to have the work done at the price the insurance company is willing to pay.

If there are injuries or a death that result from an auto accident, the auto claims adjuster may be involved in gathering medical reports and often communicates with a bodily injury adjuster, who settles medical claims.

Adjusters spend half of their time talking with policy holders, investigating claims and, in the case of outside adjusters, visiting dealerships and auto body repair shops. The other half of the adjuster's time is spent filling out the required forms and filing reports, more often than not on a computer.

An inside auto claims adjuster works primarily from an office, while an outside adjuster spends a great deal of time at various locations investigating claims. A technical specialist claims adjuster does the job of an inside or outside adjuster but is involved in more complex and technical work.

The adjuster may be aided by an auto damage appraiser, who looks over the physical damage to a car to determine the approximate cost of repairs. Based on the appraiser's cost estimates, the adjuster then places a value on the claim and negotiates a settlement.

Adjusters and appraisers work for property liability insurance companies or for independent companies that examine accidents for insurance companies.

No one is ever happy to have a car accident, but those involved can be helped through their inconvenience or distress by a claims representative who does his or her job well. If you like automobiles and also enjoy figuring out an occasional mystery, this career could suit both your interests.

What You Need to Know

❑ Basic knowledge of how insurance works
❑ Basic auto mechanics and auto body repair techniques (what's necessary and what repairs cost)

Necessary Skills

❑ Ability to negotiate a settlement that makes business sense for your company and satisfies the policyholder
❑ Basic computer know-how a plus (most adjusters now use computers to do their paperwork and to communicate with their main office)
❑ Notetaking know-how (you need to accurately record everything witnesses or policyholders tell you)
❑ Ability to ask questions, listen carefully and put together what you observe with what you hear from the policyholder

Do You Have What It Takes?

❑ A sense of objectivity that allows you to make professional judgments, regardless of the details of the accident or the persuasiveness of the policyholder or auto repair specialists
❑ Ability to remain calm when policyholders or repair specialists become difficult
❑ A good eye; you need to carefully observe damage

Education

High school diploma required; an associate degree is a big plus.

Licenses Required

Companies provide new claims adjusters and examiners with on-the-job training and home-study courses or send them to courses offered by the Insurance Institute of America, a nonprofit organization offering educational programs to professionals in the field. Their courses prepare adjusters and examiners for the licensing tests required in many states. To apply for a license you must do one or more of the following: pass a written exam; complete approved coursework; furnish character references; be at least 20 years of age and a resident of the state in which you're working.

Job Outlook

Jobs openings will grow: slightly faster than average.
The strongest demand is for auto claims representatives who know auto body shop procedures.

The Ground Floor

Entry-level job: Auto claims assistant
This clerical job can help prepare you to become a full-fledged adjuster.

Not everyone starts as an auto claims assistant; people with auto body repair experience may start as full-fledged adjusters.

On-the-Job Responsi-bilities

Beginners (auto claims assistant)

❏ Answer initial telephone calls from policyholders on how to file a claim
❏ Take information on new claims and open a file for the adjuster
❏ Make appointments for an inspection by the adjuster
❏ Keep policyholders informed on the status of pending claims

Senior Assistants

❏ Do all the above, plus handle minor claims

Experienced Adjusters

❏ Investigate claims by interviewing people involved in an accident and any witnesses
❏ Examine police reports of the accident
❏ Estimate amount of damage to the car

❏ Determine who is at fault in the accident
❏ Decide if the insurance policy covers the claim
❏ Negotiate a settlement on the claim with the policy-holder
❏ Negotiate with adjusters from other insurance companies on claims that involve their policyholders
❏ Authorize payment on the claim

Some insurance companies also hire auto damage appraisers, whose main responsibility is to determine the approximate cost of repairs.

Auto claims representatives generally work nine to five, Monday through Friday. Some adjusters have the flexibility to arrange their work schedules to meet policyholders on evenings or weekends. Some insurance companies offer or are experimenting with work-at-home opportunities, flex-time and part-time positions.

◆ **When You'll Work**

All insurance companies offer one to three weeks of paid vacations as well as holidays off.

◆ **Time Off**

❏ Pension plan
❏ Tuition reimbursement for coursework related to the job
❏ Company car or payment for use of personal car. Adjusters who work for the insurance companies owned by General Motors, Ford Motor Company and Chrysler Corporation receive company cars.

◆ **Perks**

❏ Auto insurance companies
❏ Independent insurance appraisers

◆ **Who's Hiring**

Beginners: No travel potential
Experienced outside adjusters: Some local travel. They travel to conduct investigations and visit auto body shops. How much time is spent traveling depends on

◆ **Places You'll Go**

the size of their assigned territory and the number of adjusters working in that area. An adjuster in a rural area may be required to drive longer distances because the area covers many square miles. An adjuster in a large claims office in an urban area may only drive to a small section of the city because the work is shared by a number of adjusters and the territory is small, though heavily populated.

Surroundings

Auto claims representatives work in pleasant offices. Outside claims adjusters spend up to half of their time away from the office in varied locales, such as tow yards, auto body repair shops and car dealerships.

Dollars and Cents

Pay depends on the type of job. Average annual starting salaries are as follows:
Clerk and secretaries—$15,000 to $16,400
Inside adjusters—$22,300
Senior inside adjusters—$26,700
Outside adjusters—$24,800
Senior outside adjusters—$33,200
Technical specialist claims adjusters—$39,800

Moving Up

With experience and additional on-the-job training or home-study coursework in insurance, an auto claims representative has excellent potential for advancing to jobs such as claims supervisor and claims examiner.

The claims supervisor is responsible for the investigation and settlement of all claims assigned to the department. Some offices also have a claims manager who is a level above supervisor. A claims examiner manages the settlement of complex, high-value claims and gives technical direction to adjusters. A well-established claims adjuster also can transfer from the auto field to handling marine or industrial claims or those brought on by a natural disaster.

In some companies a four-year degree is helpful for advancing into managerial positions.

Jobs are available in nearly every area of the country, although they are more plentiful in major metropolitan areas where there is a concentration of cars, accidents and claims offices.

◆ **Where the Jobs Are**

Insurance companies often provide on-the-job classroom training in company procedures and policies. Many pay for courses provided by the Insurance Institute of America, a nonprofit organization offering educational programs and professional certification. The Institute offers an Associate in Claims designation after a person successfully completes four areas of coursework and examinations in various types of claims adjusting.

◆ **Training**

Participants can prepare for the exams through independent home study, courses provided through the company or classes held at a public location such as a college campus or public meeting room.

The field of auto claims is open equally to men and women, but it is more populated by men.

◆ **The Male/Female Equation**

The Bad News

❏ Substantial paperwork
❏ Lack of variety in job tasks for inside adjusters
❏ Some companies set high quotas of cases to be handled
❏ Clients can be difficult, and difficult to please

The Good News

❏ Regular working hours
❏ Good pay, above-average benefits
❏ Potential for advancement with additional training and education
❏ Job security; industry unaffected by ups and downs of the economy

◆ **Making Your Decision: What to Consider**

More Information Please

To receive a brochure on careers in the property and casualty insurance industry, write or call:

The Insurance Information Institute
110 William Street
New York, New York 10038
212-669-9200

Alliance of American Insurers
1501 Woodfield Road, Suite 400 West
Schaumburg, Illinois 60173-4980
703-330-8500

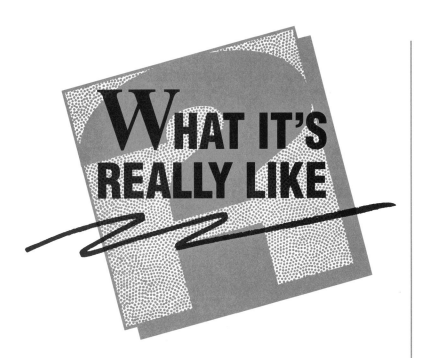

WHAT IT'S REALLY LIKE

John Parent, 28,
senior claims adjuster,
Motors Insurance Corporation,
the insurance subsidiary of
General Motors Corporation,
Detroit, Michigan
Years in the business: five

Were you always interested in the automobile industry?
I've been monkeying around with cars since I was real
young. I rebuilt and sold cars before I could drive them. I
used to hang out with one of my dad's friends who had a
small body shop. I mostly got in the way but I picked up a
lot of knowledge. My dad was supervisor for a General
Motors plant in Saginaw for 35 years. Now my brother is a
supervisor.

**Did you know early on that you wanted to go into the
automotive industry?**
At the time I graduated from high school I knew I wanted
to get into the automotive field but didn't know exactly
what I wanted to do. So I took a year off from school. I
had been working at a gas station part time and doing car
repairs at my home. I figured I'd get state certified in all

eight different mechanical fields. Initially, I was going to go to University Center, a community college, just to take the part of the program I needed to get certified. But I decided to get an associate degree in automotive service technology. We did a lot of hands-on working with cars, both body and mechanical work. I figured it could help me in the future. I went through the summer to get done as fast as I could.

How did you get your first job?
The school had a program in which you worked in a General Motors dealership for two weeks. The dealership service director or manager then submitted a grade for you to the school. After three days, the dealership hired me. I was still going to school full time and I was working at the dealership full time.

What did you do at the dealership?
Even though I got certification in all eight fields (these include tune-ups, brake and front-end work, electrical system repair, heating/air conditioning), I specialized in a couple. The car is so complicated nowadays that you can't be a genius in all eight fields. I specialized in electrical diagnosis and repair and air conditioning. Electronics always fascinated me, and it's the area where you can make the most money because everybody's afraid of it. You also don't have to get your hands dirty very much.

How long did you work at the dealership?
About two years. The warranty claims manager had left the dealership to take a job as a staff adjuster for Motors Insurance Corp. He told me about a job opening and I got hired in January 1987 as a staff adjuster.

Was that a typical job for the insurance field?
Usually Motors Insurance Corp. hires you as a claim representative. You usually spend about a year at the drive-in facility. People bring in cars involved in minor collisions. You get the general information from the customer and learn the ropes from the other adjusters. But I skipped that step. Other insurance companies use appraisers. They look at the car and write up an estimate. An adjuster handles the rest. We run the whole show from start to finish, from writing up the investigation to obtaining the check for the customer.

What is a typical day or week like on the job?

I have my office in my house. We used to write estimates by hand, but about a year ago we went to using laptop computers. In the evening I receive a fax from the main office with the list of claims that I will do the next day. After 7 P.M. I can retrieve by computer all of the information on those claims from my electronic mailbox. Sometimes I make phone calls in the evening to set up appointments. If a car is at a car dealership or tow yard, I look at the vehicle the next day and write up the estimate before I call the customer.

Do you spend a lot of time on the road?

I might only drive 30 miles a day and see three cars. On another day, I might drive 350 miles roundtrip to see one car. I spend about half my time on the road and looking at the cars. The other half I spend doing paperwork.

How is your job today as a senior adjuster different from the staff adjuster job you started in?

I got promoted to senior adjuster after almost three years. There are three major differences: more money, a nicer car and the responsibility of training new people. The new people ride with me and watch how I do the job. I keep a daily log of their progress. Then I let them start doing their own work and I critique it.

What do you like about your job?

It gives me the freedom to run my own show. As long as you do a good job, you can advance fast. I like the variety. There's something new—a different challenge and a different claim—every day. You are dealing with all types of losses, from someone hitting an animal to a collision between two cars that results in a total loss of both vehicles.

What do you dislike about the job?

Not much. I really love it; this job was made for me. I make a salary. Sometimes I might put in a 12-hour day, but some other days I might only have to put in 6 hours, so it all evens out.

What is your next step up the advancement ladder?

I want to be a claims supervisor in the field. In that job I would supervise eight to ten adjusters. I'd rotate riding with each adjuster a couple days a week to make sure they were estimating properly and doing things the way they are supposed to. I'd be there to help them with problems.

What is your proudest achievement in this field?
I'm proud of the fact that I made senior adjuster in such a short period of time. I have been told it can take seven to ten years to move from staff adjuster to senior adjuster. That's the great thing about this company; promotions are based on performance, not years of service.

What advice would you give to someone interested in being an insurance claims representative?
An associate degree in automotive technology and mechanical background are definitely important. You have to have a strong, hands-on mechanical background. If you don't know anything about a car and how to repair it, then how can you disagree with the dealership or repair shop and claim that they are charging too much? You have to be an authority on the subject.

Wayne Nagai, 41, physical damage appraiser, USAA, Seattle, Washington
Years in the business: 13

Were you always interested in cars?
As a kid I was interested in cars and hot rodding and drag racing. I was interested in the mechanics. I worked at a service station for many years through high school and then after.

What preparation did you have for your current job?
I went to the University of Washington for a year to major in engineering, but I lost interest. After spending some time in the Army Reserve, I went to work at the service station again. I heard about the auto body program at Seattle Community College and took the course.

For more than ten years I worked at a variety of jobs in body repair for dealerships and shops. I started out as an apprentice painter, then moved on to body repair and painting. I worked as manager of the body repair shop for a dealership for two years; then I was offered the job of shop manager at a Ford dealership. I worked there for a year.

How did you move into the insurance business?

Someone from USAA came into the shop and said they needed an appraiser. I said, "If you're looking for someone, why not interview me?" I got the interview and the job. I was interested because I knew I could work my own hours. If there are a lot of appraisals to do, you're busy. If business is slow, you're not. You're not tied to office hours.

What do you currently do?

I appraise damaged cars and assess what the repairs should cost. If a car is a total loss, I determine its value. Several times a day I check my phone messages. I have an office I can work from, but most of the time I work out of my car. I have a car phone and laptop computer along with all of my books. I can pull my assignments through the computer system. The computer shows me the owner's name, where the car is, the type of car and a brief description of what happened. I may receive enough assignments to keep me busy for that day or enough for a week.

I make calls to verify that the cars are where the computer says they are. I go to the locations to do damage appraisals and issue payment to the shops if I have the okay to do so. I may have to do a re-inspection. For example, if a car is in the repair shop and they find additional repairs are needed, I have to inspect it again. They may take off a bumper and find hidden damage that I couldn't see in the original appraisal. I keep checking my messages throughout the day. Contact with the car owners and the shop people is important.

How much travel is involved?

I average about 3,000 miles a month. My territory is an area of about 25 by 50 miles. Some days we run a drive-in claims center—people whose damaged cars can still be driven bring them in. We write appraisals and issue payments on the spot. We do as many as 15 cars in a day there.

How does your background in auto body repair help you as an insurance appraiser?
I know what to look for. If a fender is wrinkled, I know from experience that the part bolted inside is also damaged. My experience also puts me on better footing with body repair shop managers. They realize I've done hands-on work so they have more respect for me.

Are negotiations a major part of your job?
I have to negotiate with the auto body repair shops as well as the insured person and any third party who has a claim against USAA. If the car is totaled, its value has to be negotiated with the owner. Sometimes negotiations are necessary when USAA wants to use nonoriginal equipment or used parts, and the car owners balk at this. The hardest negotiations are with the shops on repairs and with the owners on total losses.

What do you like most about your job?
I like the fact that when it's busy, I'm working. But when it's not, I'm free to do what I want because my office is in my home. Another plus is the benefits an insurance company provides compared with those given by a small business. We have retirement benefits and there is room for advancement.

What do you like least?
I don't like all the regulations imposed on us from the insurance commission. I also don't like negotiating with body shops whose attitude is a nonnegotiable one. They want X amount of dollars to fix a car and that's it. Some shops are hard to deal with.

What is your advice to someone considering entering this field?
You should be mechanically inclined, and you should be computer literate—that's going to be necessary in the future for everybody in this line of work. I'm a computer hobbyist at home, so I learned computers inside and out on my own. I'd also recommend getting a four-year degree; it's helpful for promotion purposes.

What is your proudest achievement?
When USAA started having their appraisers use car phones

and laptop computers, I was the original tester for the new system in the Northwest. There were only seven testers in the nation, so that was an honor for me. I was given an award for helping to solve problems that were occurring during the test.

Becky Miller, 44,
auto claims adjuster,
Progressive Insurance Co.,
Doylestown, Ohio
Years in the business: eight

How did you break into the field?
I started with Progressive in 1984 as a claims assistant. I did that for two years and advanced to senior claims assistant, which I did for four years. Two years ago I became an auto claims adjuster.

What kind of preparation did you have?
I had a year of secretarial school. I went to college for a couple of years and dropped out. In 1987 I went back to college, and I've just completed a bachelor's degree in business management. To become an adjuster I had six weeks of intensive training on how to write up an estimate of damage on a car and policy interpretation. I just finished another advanced training course in physical damage.

How do you spend a typical day?
I'm in early to catch up on my mail. My hours are 7 A.M. to 3:30 P.M., though sometimes it runs over. I have a claim assigned to me, then I start scheduling appointments. I spend about half my time in the office and half out.

How do you go about doing an investigation?
I send for the police report and see if there are any witnesses. If there are, I get their statements. Then I talk with the people involved in the accident. I see if their policy covers the damage or injury. I look at the car and write up an estimate of the damage, and I make arrangements for a rental car if they need one. I send for their medical bills. If it's a cut-and-dried case, I can pay on their claim at that point.

Do you also look out for fraud?

Yes. You have to consider that whenever you're taking recorded statements. You check to see if they were ever in an accident before. You become suspicious if they appear to have prior knowledge of how to do a claim or if they have previous injuries of this type.

Does fraud occur often?

No. I had one claim in which there were two passengers in the car. When I met with the driver, he was in a big hurry to get it resolved. I was "green" so I quickly settled the claims within the same day for the three people. It was an economical settlement for Progressive; I was pleased. A month later I got information indicating that all three people had claims backgrounds. They had lied to me in the interviews. They said they hadn't filed any claims before, but they filed claims every three months. They made a career of it.

What personality traits does someone need to be an adjuster?

You have to have thick skin. People yell and scream at you a lot. Most people feel that the accident is your fault because you may be the first person they see after the accident. You have to be able to handle that. You have to be aggressive and stand your ground when you're trying to negotiate with body shops on repairs because your cost estimates don't always coincide with theirs. You have to be sympathetic but you can't go overboard. You are representing the person *and* the company. The company's goal is to earn a profit. You have to judge fairly. You also have to be aggressive in your questioning; you can't be afraid to ask things. For example, I always felt what a person makes is none of my business, but you have to ask that so you can document lost wages from an injury.

What do you like most about your work?

I like the variety. No day is ever the same. You do the same basic things but the people are always different. The job fits my personality. I like to get out in the field, drive, meet a lot of people and do something challenging. Writing estimates is challenging. There's always something you could overlook, so you have to make sure you get everything.

What do you like least?

I don't like it when the people are very upset and blame me and I can't get through to them that it's not my fault. I'm trying to help.

What is your advice to someone interested in entering this field?

Make sure you get good training. Most companies provide it. You need a well-rounded background. If you don't have a four-year degree, you should be working on one so you can be promoted. If you don't have good people skills, if you can't step back and remove yourself from a volatile situation, then you shouldn't go into it.

Mario Andretti. A. J. Foyt. Al Unser. Richard Petty. If these legends of car racing are your heroes, you might want to consider whether you, too, could literally "live in the fast lane." No driver simply jumps into an Indy race car, of course. To break in takes a willingness to invest a great deal of time and money traveling the small-track circuit to chase the dream of the big time.

For many people racing is the most exciting aspect of the automobile business. The biggest names in race car driving are among the highest-paid athletes in the world—only world champion boxers make more. The drivers' earnings are high because the dangers are high, too. The risk of injury—or death—from a crash during a 220-mph race is always present.

Most of the auto racing legends and today's rookies broke into auto racing as a hobby. They started by building their own race cars and taking them to the local dirt track in their hometowns. Then they began to travel from small track to small track. Those who won moved up to bigger name racing circuits and attracted the attention of sponsors, who provided them with parts and the financial support to keep racing. Ultimately the best drivers, the best teams and the best cars meet at races like the Indianapolis 500 and the Daytona 500 to determine the best of the best.

Racing has grown into an integral part of the automobile business, and the job of a race car driver has evolved into significantly more than climbing into a car and driving the fastest lap around the track. To attract and keep their corporate sponsors—who can spend millions to keep race car drivers on the circuit—drivers must make promotional appearances at race tracks and store openings or attend charitable events on behalf of the sponsor.

The automobile manufacturers also ask drivers to test new vehicles or a new piece of equipment. Racing allows the car companies to try the newest technology, parts and materials on the race track—improvements that eventually find their way into cars on the neighborhood dealer's showroom floor.

Although car racing is best known, drivers will race anything with wheels: motorcycles, dune buggies, tractors or trucks.

Ladies and gentlemen, start your engines.

What You Need to Know

❑ Mechanics of a car (you don't have to be able to re-build an engine, but you have to be able to pick up quickly on mechanical problems and describe them to mechanics)
❑ How to drive on a race course

Do You Have What It Takes?

❑ Competitive spirit and will to win
❑ Good communication skills (to talk with the pit crew, sponsors and the public)
❑ Discipline to regularly practice your sport
❑ Ability to concentrate and shut out all distractions when driving
❑ Courage and willingness to take risks

Physical Qualities

❑ Lean build (race cars are custom built, but the driver has to fit into a snug space)
❑ Stamina (to last long distances at high speeds, often under extreme heat conditions)
❑ Good physical condition (most drivers follow some kind of exercise regimen)
❑ Excellent hand-eye coordination
❑ Quick reflexes

Education

No formal training required

Licenses Required

Driver's license. Racers also must be licensed by the organization sponsoring the particular racing circuit, such as NASCAR (National Association for Stock Car Auto Racing, Inc.), CART (Championship Auto Racing Teams) or SCCA (Sports Car Club of America).

◆ **Getting into the Field**

Job Outlook ◆ **Competition for jobs:** extremely high.

Race car driving is more an avocation than a job for many because competing, even in amateur races, requires money.

The Ground Floor ◆ **Entry-level job:** amateur race car driver

An amateur race car driver competes at small town race tracks and eventually moves up to larger race circuits where she or he competes as an amateur with professional racers.

On-the-Job Responsi-bilities ◆ *Beginners*

❏ Build your own race car or hire people to do it
❏ Make repairs, adjustments to the race car
❏ Travel the racing circuit and compete as frequently as every week or as little as ten times a year
❏ Work another job to finance your racing passion
❏ Develop a marketing plan that shows a potential sponsor how it can use car racing to promote its products
❏ Meet with executives of potential sponsor companies to try to sell them on your idea

Experienced Drivers

❏ Work with staff of mechanics on building a race car or improving it
❏ Practice to qualify before a race
❏ Race at major events, from one a week, nearly year-round, to only a dozen events for some speciality races such as stadium truck racing
❏ Give interviews to the media
❏ Attend promotional functions as required by sponsors to make a formal speech, talk to customers and deal-ers or sign autographs

When You'll Work ◆ The racing season generally starts in February and ends around Thanksgiving. Drivers always work weekends, as all races take place on Saturdays or Sundays. Drivers spend the week leading up to a race working with their mechanics to improve the car's performance and practice driving it at top speeds to qualify for the event. Hours during the peak of the season run from early morning through evening.

Drivers take vacation time in the winter months; most racing circuits take a break during November, December and/or January.

◆ **Time Off**

For the successful:
❑ Potential for fame
❑ Opportunities for lucrative business and advertising contracts
❑ Everyday use of a car made by the sponsor
❑ Prizes in addition to cash for winning races, such as a Rolex watch for winning the Indianapolis 500

◆ **Perks**

A race car driver is an independent businessperson. As a driver succeeds on a circuit, he or she attracts more sponsors and increasing financial support.

◆ **Who's Hiring**

Beginners or experienced drivers: constant travel
Race car drivers are always on the road throughout the season. Where they go depends on the racing series. Some series, like NASCAR, are strictly domestic. Formula One racing is international, taking place mainly in Europe, but also in North and South America.

◆ **Places You'll Go**

A driver spends much of his or her time strapped inside a cramped race car. Temperatures inside the car can be over 100°F. The driver must wear a uniform and mask made of fire-retardant materials and a helmet. The rest of the time drivers are usually in the hustle-bustle environment of the garages or pits where cars are worked on. They are noisy places with engines roaring and fuel fumes in the air. Hazardous substances, including highly flammable materials like gasoline and oil, also are always present.

◆ **Surroundings**

Dollars and Cents

Professional race car drivers who keep winning can make millions of dollars. At the other end of the spectrum, amateurs spend a lot of money to keep their cars on the race circuit but may earn nothing. If an amateur is successful, he or she may receive tires or parts from a manufacturer or local distributor. As a driver moves up to higher circuits, he or she could win as little as $500 or as much as $5,000 per race.

Moving Up

Race car drivers with good business sense (or a good business manager) usually buy or become partners in related companies such as automobile dealerships. Because of their contacts with big corporate sponsors, race car drivers have great opportunities for business ventures. The biggest names get lucrative contracts to endorse commercial products.

Where the Jobs Are

Amateur and professional race tracks are located across the U.S. The highest concentrations are in the southeastern states, California and the Northeast.

Training

Race car driving schools, run by former and present race car drivers, can be helpful. There are a number in the United States and some in Europe. Sessions run full time for a minimum of one week. They are advertised in the classified sections of popular automobile magazines.

The Male/Female Equation

The overwhelming majority of race car drivers are men. Only two women—Janet Guthrie and Lyn St. James—have ever raced at the Indianapolis 500. The few women who do break in find it more difficult than men to obtain sponsors.

The Bad News

❏ Difficult to break in
❏ Extremely expensive without sponsorship
❏ Risk of injury or death
❏ Grueling schedule during race season

The Good News

❏ Potential for very high earnings
❏ Possibility of worldwide fame
❏ Extensive business opportunities if you become successful
❏ Potential for world travel

◆ **Making Your Decision: What to Consider**

◆ **Professional Organizations**

Sports Car Club of America
9033 East Easter Place
Englewood, Colorado 80112
303-694-7222

Membership of more than 53,000 auto racing enthusiasts. Sanctions, operates and administers more than 2,000 automobile events a year. Establishes competition regulations and issues competition drivers' licenses.

National Association for Stock Car Auto Racing, Inc. (NASCAR)
P.O. Box K
Daytona Beach, Florida 32015
904-253-0611

Membership of 20,000. Sanctions stock car racing events throughout the U.S. and Canada.

Championship Auto Racing Teams
390 Enterprise Court
Bloomfield Hills, Michigan 48302
313-334-8500

Sanctioning body for Indy car racing events in the United States.

WHAT IT'S REALLY LIKE

Chad DiMarco, 32,
PRO Rally race car driver
and business owner,
Huntington Beach, California
Years in the business: ten

How did you break into race car driving?
My father-in-law introduced me to it. He was a racing fanatic, and he raced himself. In 1979 we fixed up an old car and I did one race with it during college. After that race we parked the car in my parents' garage. We figured maybe there would be time to go back to it, but during college I couldn't afford to race.

In 1984 we put another car together and I raced as an amateur. At that point we did several national and professional races on the PRO Rally Circuit. I started getting some notice.

Was your goal to be a race car driver?
I didn't set out to be one. I went to college and majored in biochemistry. I was studying to be a doctor—I'm just a few courses away from a degree—but I had to quit in my senior year because of deaths in my family. I went to work in investments and did amateur racing for fun. Then someone

told me I had the makings to do very well. My father-in-law and I made proposals to several sponsors. Subaru sponsored me on a limited basis at first, but that in itself wasn't enough to support the racing. So we created a service and repair business to maintain Subaru vehicles. The racing and the repair business spawned a mail-order catalog business for safety devices, seats, whatever, for racing. That opportunity came through my relationships with the sponsors. Their financial support has increased as my relationship with Subaru enters its seventh season.

Roger Penske (owner of numerous businesses and an Indy race car team) has an adage I've adopted: you don't make money from racing, you make money from the opportunities racing opens up for you. We've created businesses to support racing.

What natural abilities helped you in racing?
I'm mechanically inclined, although I don't have any formal instruction in auto mechanics. In race car driving, you have to have natural athletic ability. You need good hand-eye coordination, clear thinking and good reaction time.

When did you feel you were established?
I'm just starting to feel that way. You start feeling established in racing when you start winning. But you don't feel like you're established in the racing *business* until the outside world starts to take notice, when you make a name for yourself. That's just happened for me. Advertising has helped with that.

How do you spend an average week or month?
I split my time among the three businesses and racing. We do ten races a year, so we travel about seven days a month. We have obligations to fulfill with our sponsors. I used to do everything, from welding the cars to finding the sponsors, to racing the car. As we became more sophisticated and Subaru became more involved, I've delegated responsibilities. Now we have mechanics. And we have a person who handles publicity and sponsorship.

What do you like most about your work?
I like the race car driving and the business equally. I enjoy winning races, but I also enjoy organizing and preparing

for the race. When a plan comes together, it's as enjoyable as winning.

What do you like least?
It's a very cut-throat business. As in all professional sports, there are a lot of applicants trying, and only a few make it. And even once you've made a name, you can't rest. You have to strive constantly to meet new goals.

What is your proudest achievement?
My first attempt at Pikes Peak Hillclimb in 1986. On qualifying day my team and I broke the qualification record of the previous year. I was named "Rookie of the Year." That was the biggest accomplishment. Also, my team had been class champion for several years, but this past season we became the overall champion with the fastest vehicle in the United States.

What is your advice to someone who would like to be a race car driver?
Make your own way. Don't follow someone else's path. You can't knock on the door of the top team and expect to break in. You have to work your way up, gain recognition and create your own alliances.

Rick Johnson, 28,
truck racer, former professional motorcycle racer, Encinitas, California
Years in the business: 19

How did you start in racing?
I started riding motorcycles when I was three. The first time I raced, I was seven. I stopped racing for a while and started again when I was nine. I thought I was going to win everything then—I came in last place. But I enjoyed myself so much I kept racing. Within a few months, I started winning.

Aren't you the winningest rider in Supercross history?
Yes. I had multiple 250, 500 and Supercross National Championships. I've had a number of international wins. I

won the MXA Rider of the Year Award three times and was on the U.S. team at the Motocross des Nationes in 1984, 1986, 1987 and 1988.

When did you feel successful?
When I won my first national event in my rookie year, 1981. But it was never enough. I used to think if I could get free parts from a store, I would be successful. Then I wanted my name in the paper. Then my picture in the paper, then a color picture in the paper. Then a picture in a clothing ad and then on the cover of the magazine. I said if I won the national championship, that would be good. Then I wanted two championships. Then I felt I had the potential to set a record. They say in the business that you are only as good as your last race. It's a never-ending battle to prove yourself.

Why did you retire from motorcycle racing at age 27?
I had a series of injuries. I had to have surgery on my right wrist. That's my throttle hand; it's the important one. I couldn't race the way I knew I was otherwise capable of doing, so I was forced to retire. It's rare to see a motorcycle racer in his thirties. The physical stress and schedule are too demanding.

So now I'm truck racing. I just signed with Mickey Thompson Entertainment Group to do offroad Grand Prix racing with trucks. It's similar to Supercross. We run races in nine stadiums, mostly on the West Coast. The course has a series of turns, jumps and bumps.

How is racing motorcycles the same or different from racing cars and trucks?
Racing is all the same. It's a matter of strategy. You try to put yourself in the right place at the right time and outthink your competitors. However, Motocross is so much more physical. Your body accounts for 80 percent; the rest is the motorcycle. In car and truck racing, the driver accounts for maybe 40 percent. Once you're strapped in, there's only so much you can do—you can only control the throttle, brake and steering wheel. You can't use your body like you can on a motorcycle.

What are your goals for truck racing?
I want to be the big fish in this pond, then I'll jump to a

bigger pond. In Motocross, I wanted to be known as the all-time, best-ever racer.

Do you want to race cars?
Yes. It's a bigger spectrum with more visibility and more money. I raced with the Barber-Saab team in the Del Mar (California) Grand Prix in 1989.

How do you handle the business side of racing?
I like to do my own business. I have a financial advisor, but I like to talk to the powers that be. If I send a lawyer to talk to Chevrolet (my sponsor), then Chevrolet will send their lawyer to talk to mine. I want it to be a relationship with me and Chevrolet, not my manager and Chevrolet.

Racers get into racing because they love it. But racing exists to sell products, whether it's a car, detergent or beer. I realized that early on in Motocross. So I tried to be the best there ever was, be it on the track or off.

What types of things do you do for your sponsors?
For Honda, I made dealer appearances. I'd go into a dealership on the Friday before a race and sign autographed pictures. For helmet, clothing and tire sponsors, I went to trade shows and signed autographs. I did a lot of testing for sponsors on new motorcycles and new tires.

What is your schedule like?
I spend every waking moment of every day on racing. In Motocross, a typical week would be Monday through Thursday in physical training—alternating between running, swimming, lifting weights—to get ready for the Supercross race on Saturday night. Then I'd do some public relations things. The season started in January and ended in November. Some of us racers then went to Europe and Japan. I basically had two weekends off at the end of the year. With truck racing I have nine races this year, about a race a month. I have several days of testing, and I focus on a triathlon training schedule.

What is your proudest achievement?
In racing there are so many emotional highs. I had one race when everything went right. It was in 1987, the last race of the season. I had won the 1986 Supercross championship but lost in 1987 due to a string of bad luck and mistakes. In

the last race, I got knocked down on the first turn but came back to win the race. I had the crowd with me. Even though I didn't win the championship, I won the race. I looked better that night than the guy who won the championship. Of course, he went to the bank the next day looking better than me.

What do you like best about racing?
It gives you a chance to express yourself like nothing else does. It's you against everyone else, especially in Moto-cross. If you did your homework and the machines run well, you take all the glory.

What do you like least?
I hate the travel. I traveled from the time I was 16 until I was 27 and didn't slow down. I didn't have time to enjoy my home and personal life. The irony is that now with truck racing, I'm traveling a lot less but want to race more.

What is your advice to someone who wants to go into racing?
Look at all the angles before you jump in. If you don't have the support of family, friends and other people to help you financially, it's hard to pursue it aggressively because it is so expensive.

Robbie Buhl, 27,
race car driver, Firestone Indy Light Series, Detroit, Michigan
Years in the business: seven

What made you decide to be a race car driver?
I went to the Indy 500 when I was about nine years old. It consumed me. I knew then that I definitely wanted to race cars. When I was 13 or 14, I raced against my three brothers in go-karts and on minibikes. I did a lot of motorcycle racing.

How did you get into car racing?
While I was in England, I went to a racing school—the Jim Russell school. I did a couple of amateur races in England in Formula Fords, and that reconfirmed my boyhood

dreams. It was the end of the season and there were only a couple of races left. I won both of them. I knew then I wanted to continue.

It took two years after that to get back to racing. I went to the Detroit Auto Show in 1985. They had a Sports Renault on display that cost $9,999 with everything included. There was a pro racing series for the car. Everybody had the same car, the driver was the only variable. I thought that was perfect for me. It took some time to talk my parents into letting me buy the car with some of the money they had set aside for me.

I didn't know anything at first about the car, but I learned. I found some guys to answer my questions and had one guy who helped me go through the car every couple weeks. I still didn't know much about the sport. I only knew about Indy cars—I didn't know there were hundreds of other racing series below that. When I found out about them, I went to as many tracks as I could by myself, towing the car behind me. That helped me learn about the sport. I was always in the top three, and I won the East Coast regional championship.

Did you move on to other racing series?
I did the Barber-Saab pro series in 1987. That was my big test to see if I really was competitive enough or not. In the two races I took the pole position (the inside front spot), and I led the race until my car broke down. It was a disappointment, but I proved to myself I could be as quick or quicker than the other drivers. I was happy to know I could be competitive. Financially those races hurt. Had I won both, I would have taken home $12,000. Instead it ended up costing me money. I asked the series to work with me on the money end. They did, and I went to five more races. My best finish was a second, and I ended up seventh in the series.

What is your proudest achievement?
It was a big achievement when I jumped from Sports Renault to the Barber-Saab series and was instantly competitive. When I went on to win the Barber-Saab series in 1989 by winning more races than anybody had ever won before, that was a great achievement. The fact that I am in the Fire-

stone Indy Lights Series, have stayed competitive and have gotten sponsors year after year is an achievement. When Ford Motor Co. called me to drive for their Roush team this year, that was an achievement. When people call you to drive their car, that's making it in the sport.

What do you like most about car racing?
Everybody says the speed, but for me it's just being the quickest on that given weekend. It's a good feeling inside when you are the man setting the pace. It's the challenge of getting the car so it's the quickest one out there. It's the challenge of racing through the corners with another car. It's the challenge of making a perfect lap.

What do you like least?
I'm tired of working so hard at it away from the track. Even though I have the best sponsors I've ever had, I'm still in debt from earlier racing. It's a scramble each day to keep everything covered. And it prevents me from doing what I really want to do—driving the race car. I'd like to concentrate on driving.

What else do you have to do?
It's no longer a matter of XYZ Widget Co. putting their name on the car and you verbally guaranteeing them a return on their investment. The driver has to do a lot of things away from the track to make the sponsors' return on their investment. The company has to create a whole marketing plan with the focal point being the race car. For example, during lunch time at a recent race, we took the car to a sponsor's facility. Then we took it to another sponsor's stores for a couple of hours and signed autographs.

What is the charitable work you do?
Racing for Kids is a program to raise money for children's hospitals. I got involved when it was in its idea stage. I suggested we put ''Racing for Kids'' on the race car and visit children's hospitals in the city where races are held. The program has taken off and has become a great thing. We've raised hundreds of thousands of dollars. At one race alone this year, we raised more than $60,000.

At the hospitals I give the kids a hat, picture or Hot Wheels car, for example. I take my helmet with me so that they can try it on. Some kids ask for a hug. Some are real race fans. They all love having a visitor.

What do you get out of it?
To succeed in racing sometimes you have to wear blinders. You worry if the next deal will come along and think the world will end tomorrow if it doesn't. But then you go and see these kids cruising around the hospitals with IVs hanging out of their arms, and that doesn't faze them. It's a great reality check for me. It makes me realize that in comparison I have life on a silver platter. Being involved with Racing for Kids is also a chance for me to give back something to the sport I love.

Do you feel established?
No. I'm still trying to fight my way to the top. I want to get to the next level, Indy cars. I constantly wonder if I can be competitive.

What advice do you have for someone who'd like to be a race car driver?
If it's all you've dreamed of, let the dream help you over the highs and lows of the sport. You have to be willing to give up everything else to try to make it—it is so hard and so few people can. I'm not to the top yet. And I may not get there.

WILL YOU FIT INTO THE WORLD OF CARS?

Before you enroll in a program of study or start to search for a job in one of the careers described in this book, it's smart to figure out whether that career is a good fit, given your background, skills and personality. There are a number of ways to do this. They include:

❏ Talk to people who work in the field. Find out what they like and don't like about their jobs, what kinds of people their employers hire and what their recommendations are about training.

❏ Use a computer to help you identify career options. Some of the most widely used programs are *Discover*, by the American College Testing Service, *SIGI Plus*, developed by the Educational Testing Service and *Career Options* by Peterson's. Some public libraries make this career software available to library users at low or no cost. The career-counseling or guidance offices of your high school or local community college are other possibilities.

❏ Take a vocational interest test. The most commonly used ones are the Strong-Campbell Interest Inventory and the Kuder Occupational Interest Survey. High schools and colleges usually offer free testing to their students and alumni through their guidance and career-planning offices. Many career counselors in private practice at community job centers are also trained to interpret results.

❑ Talk to a career counselor. You can find one by asking friends and colleagues if they know of any good ones. Or contact the career information office of the adult education division of a local college. Its staff and workshop leaders often do one-on-one counseling. The job information services division of major libraries sometimes offer low- or no-cost counseling by appointment. Or check the *Yellow Pages* under the heading "Vocational Guidance."

Before you spend time, energy or money doing any of the above, take one or more of the following five quizzes (one for each career described in the book). The results can help you confirm whether you really are cut out to work in a particular career.

If a career in CAD (computer-aided design) interests you, take this quiz:

Read each statement, then choose the number 0, 5 or 10. The rating scale below explains what each number means.

0 = Disagree
5 = Agree somewhat
10 = Strongly agree

____I understand how a car operates.

____I enjoy working on computers and understand how to use them

____I have good powers of concentration

____I think I am capable of working alone for long periods of time in front of a computer screen

____I am good at asking questions and following directions

____I like to identify problems and find a solution for them

____I am meticulous in terms of neatness, accuracy and attention to detail

____I have strong math skills, especially algebra, geometry and trigonometry

____I've worked with factory machines to make parts in shop class

____I have a strong background in science, including physics and chemistry

Now add up your score.   ____Total points

If your total points were less than 50, you probably do not have sufficient interest or inclination to learn what's required to go into computer-aided design. If your total points were between 50 and 75, you may have what it takes to get into computer-aided design, but be sure to do more investigation by following the suggestions at the beginning of this section. If your total points were above 75, it's highly likely that you are a good candidate to work in the field of computer-aided design.

If vehicle sales interests you, take this quiz:

Read each statement, then choose the number 0, 5 or 10. The rating scale below explains what each number means.

0 = Disagree
5 = Agree somewhat
10 = Strongly agree

____I love talking about cars and learning about the newest automotive technology
____I am a disciplined and highly motivated person
____I have a friendly, outgoing personality
____I'm not afraid of approaching people I don't know by phone or in person
____I have sales experience
____I like explaining the details of a product to a customer
____I have a persuasive manner
____I don't take rejection personally
____I'm willing to work long hours
____I'm good at negotiating and making compromises

Now add up your score.   ____Total points

If your total points were less than 50, you probably do not have what it takes to become a successful car salesperson. If your total points were between 50 and 75, you may have what it takes to get into automobile sales, but be sure to do more investigation by following the suggestions at the beginning of this section. If your total points were above 75, it's highly likely that you are a good candidate to work in the field of automobile sales.

If a career as an auto mechanic interests you, take this quiz:

Read each statement, then choose the number 0, 5 or 10. The rating scale below explains what each number means.

0 = Disagree
5 = Agree somewhat
10 = Strongly agree

___I'm good at taking things apart and putting them back together

___I like working with my hands and am good at it

___I don't mind getting dirty

___I know how to work with tools

___I understand how a car operates

___I have no fear of using computers

___I have good reading skills

___I like figuring out where a problem exists and how to solve it

___I like to stick with a problem or task until it is completed

___I am good at asking questions

Now add up your score.   ___Total points

If your total points were less than 50, you probably do not have sufficient interest or inclination to learn what's required to become an auto mechanic. If your total points were between 50 and 75, you may have what it takes to get into auto mechanics, but be sure to do more investigation by following the suggestions at the beginning of this section. If your total points were above 75, it's highly likely that you are a good candidate to work in the field of auto mechanics.

If a career as an auto claims representative interests you, take this quiz:

Read each statement, then choose the number 0, 5 or 10. The rating scale below explains what each number means.

0 = Disagree
5 = Agree somewhat
10 = Strongly agree

____I am a good listener

____I have some understanding of or am interested in learning about auto mechanics and auto body repair

____I have a basic understanding of how insurance works

____I am open minded and able to make an objective judgment after obtaining the facts about a situation

____I'm not shy about asking people personal questions

____I am able to negotiate with people to come up with a solution that's acceptable to all

____I know the basics of using a computer

____I remain calm when others around me are upset

____I'm good at taking notes

____I am observant; I notice and remember the tiniest of details

Now add up your score.   ____Total points

If your total points were less than 50, you probably do not have sufficient interest or inclination to learn what's required to become an auto insurance claims representative. If your total points were between 50 and 75, you may have what it takes to get into auto insurance claims, but be sure to do more investigation by following the suggestions at the beginning of this section. If your total points were above 75, it's highly likely that you are a good candidate to work in the field of automobile insurance claims.

If a career as a race car driver interests you, take this quiz:

Read each statement, then choose the number 0, 5 or 10. The rating scale below explains what each number means.

0 = Disagree
5 = Agree somewhat
10 = Strongly agree

____I love to drive fast

____I understand the mechanics of a car

____I am a good driver

____I have good hand-eye coordination and strong reflexes

____I am in good physical condition

____I have strong powers of concentration

____I am a highly competitive person

____I have a good business sense and understand marketing

____I can comfortably and confidently make presentations to groups of people

____I have a great deal of physical stamina

Now add up your score.   ____Total points

If your total points were less than 50, you probably do not have what it takes to become a race car driver. If your total points were between 50 and 75, you may have what it takes to get into race car driving, but be sure to do more investigation by following the suggestions at the beginning of this section. If your total points were above 75, it's highly likely that you can compete in the field of car racing.

ABOUT
THE AUTHOR

Michelle Krebs is a Detroit-based freelance writer specializing in automotive coverage for newspapers, magazines and book publishers. She was a staff writer and financial editor for *Automotive News*, a weekly trade publication. Krebs started covering the auto industry as business editor of the *Oakland Press*, a daily newspaper in Pontiac, Michigan, a suburb of Detroit.

She is the author of *The U.S. Car Market: Prospects Through the 1990s*, published in 1991 by the Economist Intelligence Unit in London. The book is an extensive examination of what has occurred in the U.S. auto industry over the past 15 years and a forecast of what is likely to happen in the next decade.

Krebs is a writer and researcher on a number of other upcoming auto-related books. She also is a regular contributor to such magazines as *AutoWeek*, *Automobile* and *Consumer's Digest*, as well as newspapers, including the *Detroit News* and the *New York Times*.